Francisco Zayas-Gato · Antonio Díaz-Longueira ·
José-Luis Casteleiro-Roca · Esteban Jove
Editors

Distributed Computing and Artificial Intelligence, Special Sessions III - Intelligent Systems Applications, 21st International Conference

Springer

Editors
Francisco Zayas-Gato
Department of Industrial Engineering
University of A Coruña
Ferrol, Spain

Antonio Díaz-Longueira
Department of Industrial Engineering
University of A Coruña
Ferrol, Spain

José-Luis Casteleiro-Roca
Department of Industrial Engineering
University of A Coruña
Ferrol, Spain

Esteban Jove
Department of Industrial Engineering
University of A Coruña
Ferrol, Spain

ISSN 2367-3370 ISSN 2367-3389 (electronic)
Lecture Notes in Networks and Systems
ISBN 978-3-031-73909-5 ISBN 978-3-031-73910-1 (eBook)
https://doi.org/10.1007/978-3-031-73910-1

This Springer imprint is published by the registered company Springer Nature Switzerland AG
The registered company address is: Gewerbestrasse 11, 6330 Cham, Switzerland

Preface

Due to climate change, the aging of the population, and the upcoming requirements of industry and engineering, new challenges have become of more interest and need to be achieved. Thus, minimizing energy consumption, reducing air pollution and emissions of greenhouse gases, improving the quality of life for human beings, or even achieving industry requests are becoming of high interest and need to be considered by industrial and academic personnel, as they are in charge of the research and improvements that will define the future.

Consequently, the industry needs to engage with the upcoming techniques developed in building management, assistive technologies, and applied engineering to achieve the previously mentioned goals. Until now, traditional approaches have performed very efficiently in response to current demands. Nevertheless, new advanced methods are necessary for future and upcoming requests.

The Special Session on Intelligent Systems Applications (ISA) under the framework of the 21st International Symposium on Distributed Computing and Artificial Intelligence 2024 (DCAI 2024) is a significant forum for the presentation of development and applications of innovative techniques in closely related areas. The exchange of ideas between scientists and technicians from both academic and business areas is essential to facilitate the development of systems that meet the demands of today's society. Technology transfer in this field is still challenging, so this type of contribution is mainly considered in this symposium. ISA Special Session brings discussions and publications on developing innovative techniques for complex problems. This year's technical program will present high quality and diversity, with contributions in well-established and evolving research areas. The ISA Special Sessions technical program has selected ten papers for its publication. This year, ISA Special Session has teamed up with "Logic Journal of the IGPL" (Oxford University Press) for one particular issue, including selected papers from ISA Special Session 2024.

DCAI 2024 was organized by the University of Salamanca and was held from 26 to 28 June 2024 in the city of Salamanca (Spain). We would like to thank all the contributing authors, National Associations (AEPIA, APPIA), and sponsors (AIR Institute). Finally, we are grateful and value the hard work of the program committee members, which has been essential for the success of the ISA Special Session 2024.

June 2024

Francisco Zayas-Gato
Antonio Díaz-Longueira
José-Luis Casteleiro-Roca
Esteban Jove

Lecture Notes in Networks and Systems 1173

The series "Lecture Notes in Networks and Systems" publishes the latest developments in Networks and Systems—quickly, informally and with high quality. Original research reported in proceedings and post-proceedings represents the core of LNNS.

Volumes published in LNNS embrace all aspects and subfields of, as well as new challenges in, Networks and Systems.

The series contains proceedings and edited volumes in systems and networks, spanning the areas of Cyber-Physical Systems, Autonomous Systems, Sensor Networks, Control Systems, Energy Systems, Automotive Systems, Biological Systems, Vehicular Networking and Connected Vehicles, Aerospace Systems, Automation, Manufacturing, Smart Grids, Nonlinear Systems, Power Systems, Robotics, Social Systems, Economic Systems and other. Of particular value to both the contributors and the readership are the short publication timeframe and the world-wide distribution and exposure which enable both a wide and rapid dissemination of research output.

The series covers the theory, applications, and perspectives on the state of the art and future developments relevant to systems and networks, decision making, control, complex processes and related areas, as embedded in the fields of interdisciplinary and applied sciences, engineering, computer science, physics, economics, social, and life sciences, as well as the paradigms and methodologies behind them.

Indexed by SCOPUS, EI Compendex, INSPEC, WTI Frankfurt eG, zbMATH, SCImago.

All books published in the series are submitted for consideration in Web of Science.

For proposals from Asia please contact Aninda Bose (aninda.bose@springer.com).

Organization

Organizing Committee

Chairs

Francisco Zayas-Gato	University of A Coruña, Spain
Antonio Diaz-Longueira	University of A Coruña, Spain
José-Luis Casteleiro-Roca	University of A Coruña, Spain
Esteban Jove	University of A Coruña, Spain

Programme Committee

Álvaro Michelena Grandío	University of A Coruña, Spain
Andres Pinon	University of A Coruña, Spain
Carmen Benavides	University of Leon, Spain
Daniel Urda	University of Burgos, Spain
Dragan Simic	University of Novi Sad, Serbia
Hector Alaiz Moreton	University of Leon, Spain
Héctor Quintian	University of A Coruña, Spain
Isaias Garcia	University of Leon, Spain
José Aveleira Mata	Research Institute of Applied Sciences in Cybersecurity (RIASC), Spain
José Luis Calvo Rolle	University of A Coruña, Spain
Jose Manuel Gonzalez-Cava	University of La Laguna, Spain
Liliana Durón	University of Salamanca, Spain
Maria Teresa García	University of Leon, Spain
Martín Bayón Gutiérrez	University of Leon, Spain
Miriam Timiraos Díaz	University of A Coruña, Spain
Natalia Prieto Fernández	University of Leon, Spain
Paula Arcano Bea	University of A Coruña, Spain
Ramon Costa-Castelló	Polytechnic University of Catalonia, Spain
Roberto Casado-Vara	University of Burgos, Spain
Santiago Torres	University of La Laguna, Spain
Sara Rodríguez	University of Salamanca, Spain
Sérgio Ramos	Polytechnic of Porto, Portugal

DCAI Organizing Committee

Advisory Board

Kenji Matsui	Osaka Institute of Technology, Japan
Yuncheng Dong	Sichuan University, China
Francisco Herrera	University of Granada, Spain
Tan Yigitcanlar	Queensland University of Technology, Australia
Tiancheng Li	Northwestern Polytechnical University, China

Program Committee Chairs

Ravikumar C. V.	Vellore Institute of Technology University, India
Pawel Sitek	Kielce University of Technology, Poland
Nasro Min Allah	Imam Abdulrahman Bin Faisal University (KSA), Saudi Arabia

Special Sessions Chairs

Rashid Mehmood	King Abdulaziz University, Saudi Arabia
Goreti Marreiros	ISEP/GECAD, Portugal

Organizing Committee Chairs

Sascha Ossowski	King Juan Carlos University, Spain
Sara Rodríguez	University of Salamanca, Spain
Guillermo Hernández González	University of Salamanca, Spain
Laura Grande Pérez	University of Salamanca, Spain

Organizing Committee

Fernando De la Prieta (Chair)	University of Salamanca, Spain
Sara Rodríguez González	University of Salamanca, Spain
Javier Prieto Tejedor	University of Salamanca and AIR Institute, Spain
Ricardo S. Alonso Rincón	AIR Institute, Spain
Alfonso González Briones	University of Salamanca, Spain
Pablo Chamoso Santos	University of Salamanca, Spain
Javier Parra	University of Salamanca, Spain
Belén Pérez Lancho	University of Salamanca, Spain
Ana Belén Gil González	University of Salamanca, Spain
Ana De Luis Reboredo	University of Salamanca, Spain

Angélica González Arrieta	University of Salamanca, Spain
Angel Luis Sánchez Lázaro	University of Salamanca, Spain
Emilio S. Corchado Rodríguez	University of Salamanca, Spain
Guillermo Hernández	University of Salamanca, Spain
Liliana Durón	University of Salamanca, Spain
Laura Grande	University of Salamanca, Spain
Carmen Losada	AIR Institute, Spain
Albano Carrera González	AIR Institute, Spain
Marta Plaza Hernández	University of Salamanca, Spain
Beatriz Bellido	University of Salamanca, Spain
María Alonso	University of Salamanca, Spain
Diego Valdeolmillos	AIR Institute, Spain
Carlos Álvarez	AIR Institute, Spain

DCAI 2024 Sponsors

Sponsors

Organizers

Support from Research Associations, Institutions & Labs

Contents

A Machine Learning - Based System for Determining Water Potability

Míriam Timiraos[1,3(✉)], Antonio Díaz-Longueira[1(✉)], Francisco Zayas-Gato[1], José-Luis Casteleiro-Roca[1], Óscar Fontenla-Romero[2], and José Luis Calvo-Rolle[1]

[1] CTC, CITIC, Department of Industrial Engineering, University of A Coruña, Calle Mendizábal s/n, Campus de Esteiro, 15403 Ferrol, Spain
{a.diazl,f.zayas.gato,jose.luis.casteleiro,jlcalvo, miriam.timiraos.diaz}@udc.es

[2] LIDIA, CITIC, Department of Computer Sciences and Information Technologies, University of A Coruña, Calle Mendizábal s/n, Campus de Esteiro, 15403 Ferrol, Spain
oscar.fontenla@udc.es

[3] Department of Water Technologies, National Technological Center, Fundación Instituto Tecnológico de Galicia, 15003 A Coruña, Spain

Abstract. This research focuses on enhancing water potability classification through the integration of three machine learning techniques. A comparative analysis of diverse classification methods is conducted, incorporating multiple thresholds by employing a variable extraction approach. The primary objective is to streamline the input set, aiming for a significant reduction in computational costs and model complexity. This streamlined approach not only facilitates a more efficient training process but also implies a shorter duration for model training. To rigorously evaluate the model's performance, a K-fold cross-validation is implemented within this framework. This comprehensive approach contributes to the advancement of water quality assessment methodologies, with potential implications for improving the efficiency and reliability of potability water classification models.

Keywords: Water potability · Classification · Feature Extraction · Machine learning · AUC · Fit Time

1 Introduction

Water stands as the paramount source of life, sustaining countless living organisms [12]. The growth and vitality of all life on Earth hinge on access to clean and uncontaminated water for daily use. Ongoing pollution of water poses a severe threat to the organisms reliant on it, with recent studies indicating a continuous decline in water quality, rendering it less suitable for consumption [9]. Human and industrial activities, including chemical fertilizers in agriculture and land

F. Zayas-Gato et al. (Eds.): DCAI 2024, LNNS 1173, pp. 1–10, 2024.
https://doi.org/10.1007/978-3-031-73910-1_1

use alterations, are identified as key contributors to the deteriorating state of water quality [3]. The World Health Organization (WHO) expresses alarm at the challenges associated with providing safe drinking water due to escalating pollution in urban areas [13]. Ineffectual water management is recognized as a significant external factor causing various diseases [6], with the United Nations (UN) reporting approximately 1.5 million annual deaths attributed to water-borne diseases [11].

Addressing the global implications of the drinking water problem becomes imperative to prevent further escalation. Sustained efforts are necessary to preserve the quality of clean water [6]. An effective approach involves assessing both water quality and the surrounding environment. Water quality testing serves as a crucial tool to determine its suitability for human consumption. Typically, tests examine water characteristics in terms of physical, biological, and chemical conditions [11]. Various metrics gauge water quality, with the Water Quality Index (WQI) being one prominent method. While traditional testing methods are time-consuming, the urgency of the issue necessitates quicker water quality assessments. Machine learning emerges as a viable technique to expedite the process [16].

Several classification methods, such as Decision Trees, Naïve Bayes, Support Vector Machine, K-Nearest Neighbor, and K-Means, can be employed to characterize water quality. Past studies have explored machine learning applications extensively. For instance, Sulaiman et al. applied an artificial neural network (ANN) to classify water levels in Malaysia [15]. The work proposed in [12] compares SVM, KNN, and Naïve Bayes algorithms to assess water quality in India. Cumulatively, these studies underscore the potential of machine learning systems to rapidly determine water quality, offering a promising solution to this critical issue [11].

In this project, three machine learning techniques have been amalgamated to enhance the classification of water potability. Different classification methods will be compared, incorporating various thresholds through a variable extraction method. The aim is to simplify the input set, leading to a reduction in computational costs and model simplicity, thereby implying a shorter training time. A K-fold cross-validation is implemented in this function to assess the model's performance.

This paper is structured as follows: after this introduction, the dataset is described. Then, the methods used are explained. After that, the description of the experiments and results are presented. Finally, the conclusions are drawn, and future works are proposed.

2 Dataset Description

The dataset under consideration comprises a collection of water quality measurements. The main objective of this dataset is to assess the potability of water based on a diverse set of constituents. Each entry in the dataset includes measurements of different water quality parameters, covering physical, chemical, and biological attributes.

The data set is structured in a tabular format, where each row corresponds to a specific water sample, and the columns represent the diverse set of measured parameters. In addition, a specific column serves as a target variable, indicating the binary classification of whether the water is considered potable or non-potable. The total number of samples selected amounts to 34944. The dataset has been extracted from a public dataset bank, called Kaggle [1].

The variables and a brief explanation are presented in Table 1.

Table 1. Variables in the dataset under study.

Variable	Description	Type
pH	Acidity or alkalinity of the water	Float
Iron	Quantity of iron.	Float
Nitrate	Quantity of nitrate.	Float
Chloride	Quantity of chloride.	Float
Lead	Quantity of lead.	Float
Zinc	Quantity of zinc.	Float
Turbidity	Turbidity of the water.	Float
Fluoride	Quantity of fluoride.	Float
Copper	Quantity of copper.	Float
Odor	Odor.	Float
Sulfate	Quantity of sulfate.	Float
Conductivity	Quantity of conductivity.	Float
Chlorine	Quantity of chlorine.	Float
Magnesium	Quantity of magnese.	Float
Total Dissolve Solids	Total Dissolve Solids.	Float
Potability	Target variable: potability (1) or not (0).	Bool

In the last row of the table it can be seen that the objective variable has two values, 0 or 1, which indicate the potability status of the water. The sample count for each possible outlet is as follows: 16,250 cases where the water is not potable, and 18,694 cases where the water is potable. This demonstrates the presence of a balanced data set.

Using this dataset under study allows for a robust exploration of the complex interaction between different water components and their influence on water drinkability.

3 Applied Methods

Two stages are implemented to classify whether or not the provided water sample is potable. In the first one, the dimensionality of the data set is reduced, while in the second one, a set of classification methods is implemented, choosing the best one. Therefore, this section is divided into two subsections. In the Sect. 3.1 the feature extraction method employed will be described, while in the Sect. 3.2 the three different classification techniques implemented will be defined.

3.1 Feature Extraction

As indicated in Sect. 2, the operational dataset encompasses 15 variables. This quantity of attributes may entail a substantial computational burden during the training of models and introduce mathematical intricacies in classifier systems. Consequently, in such situations, utilizing methods for reducing dimensionality through feature extraction is customary. The objective is to decrease the count of variables in the dataset, consequently mitigating computational expenses and yielding streamlined classifiers that maintain satisfactory performance. Now, a diverse array of techniques and algorithms for dimensionality reduction is available, and Principal Component Analysis (PCA) stands out as one of the most frequently employed methods.

Principal Component Analysis (PCA). Principal Component Analysis (PCA) is an unsupervised multivariate statistical approach developed by Pearson [7], commonly used for dimensional reduction of datasets. This technique characterizes the variation in a multivariate dataset as a set of uncorrelated variables that correspond to linear combinations of the original parameters. The primary goal of this strategy is to generate new orthogonal axes that maximize the variance of the data while avoiding information loss. This is achieved by calculating the eigenvalues of the correlation matrix. Subsequently, the initial set can be linearly translated to a lower-dimensional space using the eigenvectors [17].

3.2 Classification Techniques

Next paragraphs briefly describe the three supervised classification techniques applied in this work.

K-Nearest Neighbors (KNN). K-Nearest Neighbors (KNN) is a popular machine learning algorithm for classification and regression. It is based on the idea that similar data shares labels or values. In the training phase, it uses the entire data set as a reference. During predictions, it calculates the distance between the input point and all training points, identifying the K nearest neighbors. In classification, it assigns the most common label among the K neighbors as a prediction, while in regression, it averages the target values of the K neighbors to predict the input point value [2,8].

Support Vector Machine (SVM). Support Vector Machines (SVM), developed by Cortes and Vapnik [5], are widely recognized supervised techniques. They seek to establish a hyperplane that maximizes the minimum distance (margin) between the hyperplane and the closest samples of each class, thus defining a limit for classifying new samples. Although SVMs assume a linear boundary, many real-world data sets exhibit nonlinear separability. To address this, SVMs use data transformations using selected kernel functions [10].

Deep Neural Network (DNN). Deep Neural Networks (DNN) belong to a category of machine learning algorithms that share similarities with artificial neural networks, aiming to emulate the information processing observed in the human brain. DNNs encompass multiple hidden layers (l) positioned between the input and output layers [14]. Each layer consists of a specific number of units (neurons) that apply a defined functional transformation to the input. These models have the capability to approximate the behavior of any function, as per the universal approximation theorem [4]. The output (y) of a unit (i) in layer (l) is linked to the output (x) of the preceding layer (k) with J outputs, utilizing a set of weights ($wi,\ k$), a bias (b), and a non-linear activation function f (1).

$$y_i^l = f \sum_{j=1}^{J} w_{i,k} x_k + b_i \quad (1)$$

4 Experiments and Results

The present section describes the setup of the experiments and the results obtained.

4.1 Experiment's Setup

This section provides the experiment setup, including the tools and metrics used to measure and compare the performance of the classification methods.

Data Preprocessing. The first step was the preprocessing of the case study data set. Once the data were analyzed, the samples with missing data were located and completed with the mean of the samples of the variable in question. Furthermore, outliers were filtered out and, finally, the data were standardized by scaling the entries to the unit variance.

Feature Extraction. Following the preparation of the dataset, the PCA method is employed to diminish the number of features and identify the most crucial components. The determination of the optimal count of principal components is based on an examination of the variance elucidated by each component, with a focus on those contributing substantially to the variance. To achieve the most effective classifier, considering both performance and computational cost efficiency, the experiments in this study were conducted while considering various numbers of components.

Classification. In order to carry out the classification phase of the incoming samples, the following two phases must be carried out.

Classifier Configuration. Each of the supervised classification techniques presented in Sect. 3.2 has been tested with different configurations. Table 2 shows the hyperparameters that have been configured for each algorithm as well as the values that have been implemented.

- KNN: for this technique, two hyperparameters have been tested. On the one hand, nearest neighbors have been evaluated for different numbers, from 1 to the square root of the total number of samples of the output column (a value that has been rounded off for testing). On the other hand, the weight function used in the prediction has been tested for uniform and distance values. The former refers to the fact that all the points in each neighborhood are weighted equally, instead of the latter weight points by the inverse of their distance., i.e., closer neighbors of a query point will have a greater influence than neighbors that are further away.
- SVM: in this case, different configurations have been tested by modifying the algorithm kernel, which indicates the transformation function. The kernels used are: rbf (Radial Basis Function), poly (polynomial), and sigmoid.
- DNN: deep neural networks have been analyzed by iterating with the number of node neurons in the two layers. For each parameter, values between 5 and 16 have been run with a step of 1. It is important to note that the ReLu function was used as the activation function in the input and hidden layer neurons and Softmax in the output layer. This configuration is commonly used in classification tasks with neural networks.

Table 2. Techniques configurations

Classification Technique	Hyperparameter configuration	Tested Values
KNN	Number of Neighbors (K)	$1:1:\sqrt{n}$
	Weights	uniform, distance
SVM	Kernel	rbf, poly, sigmoid
DNN	Number of nodes first layer	$5:1:16$
	Number of nodes second layer	$5:1:16$

* n = total number of samples of the output column

Classifier Assessment. For the training process of each model, a k-fold cross-validation with a k value of 5 has been employed. Additionally, the evaluation metric chosen is the Area Under the Receiver Operating Characteristic Curve (AUC), widely used in classification tasks. This parameter is utilized to establish the relationship between true positive and false positive rates, providing two fundamental advantages. Firstly, it offers a unified assessment of the classifier's performance, and secondly, it remains unaffected by variations in class distribution [18]. On the other hand, the computational cost of each implemented classifier has also been measured. This was achieved by considering the average training time for each model configuration. In this sense, it must be taken into account that the experiments have been executed on a computer with an Intel(R) Core(TM) i7-1360 CPU @ 2.20 GHz, NVIDIA® GeForce RTX™ 3050 Laptop GPU, and a RAM of 32 GB.

4.2 Results

This section shows the results derived from the experimental setup described above. Initially, to establish the optimal number of components for reducing the original dataset, an initial Principal Component Analysis (PCA) was performed. This analysis aimed to identify the components along with their corresponding percentages of explained variance. The outcomes are visualized in Fig. 1 in the form of a bar graph.

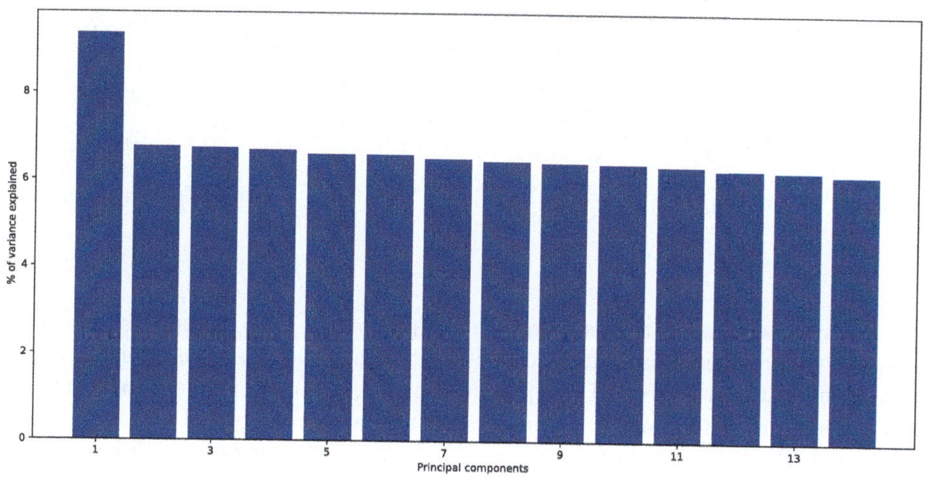

Fig. 1. PCA results.

Based on the achieved results in Fig. 1, two different component selections will be considered for the experiments for evaluating the performance of combining dimensional reduction with the above-described classification techniques in terms of classification accuracy and computational cost. The two components selection criteria are as follows:

- Components with a percentage of explained variation greater than 7%: in this case, the first component is selected.

- In the second case, the 14 components are selected given the similarity between the explainability of all the components.

After selecting the different numbers of components to be used in each experiment, the performance of each technique using the proposed configurations is tested and evaluated. Before presenting the results, since the models were fitted and evaluated by k-fold cross-validation, it is essential to underline that the mean values of AUC and training time are represented.

Table 3 contains the best results for each classification technique, indicating the configuration for each case. In addition, the results are presented according to the components extracted from the PCA method.

Table 3. Results of the three classification techniques for each PCA configuration

PCA comp.	Classif. Tech.	Best Configuration	AUC (%)	T. Time (s)
1 component	KNN	K = 180 Weights = Uniform	87.509	0.092
	SVM	Kernel = rbf	89.237	204.991
	DNN	No.of nodes first layer = 12 Num. of nodes second layer = 13	90.518	868.763
14 components	KNN	K = 180 Weights = Distance	87.713	0.150
	SVM	Kernel = rbf	89.237	1344.66
	DNN	No.of nodes first layer = 15 Num. of nodes second layer = 15	90.614	367.845

As can be seen, the performance results with 14 components extracted improve concerning those with 1 component, this is because the explained variance is similar to the second component, that is, all of them are important. On the other hand, the training time worsens noticeably in two of the techniques, KNN and SVM, leaving DNN with a significant improvement for this metric.

5 Conclusions and Future Work

This research analyzes the performance of three supervised classification techniques in combination with the PCA dimensional reduction method to indicate the potability of water.

The results obtained have been very promising, reaching AUC values higher than 89% except for the KNN method which has reached around 87% AUC. Considering only the classifier performance, DNN has been shown to detect better water potability, reaching 90.6% AUC for the network with 15 nodes in the first layer and 15 nodes in the second layer. However, the computational cost of this technique, with an average training time of 367.85 s, is significantly higher than other methods, e.g., KNN, with more than 87% AUC and a training time of 0.15 s. Therefore, considering a computational performance-cost relationship, it can be concluded that KNN is the best technique.

On the other hand, when comparing the results obtained using a different number of components, it was observed that a significant reduction in the number of components reduces the training times of the model, although a worsening of the performance of the classifier for the case studied. In the case of KNN, a maximum loss of 0.2% of the AUC is quite optimal compared to the substantial reduction of the training time. This aspect is similarly reflected in SVM, where component reduction greatly reduces the computational cost while keeping the classifier performance at the same value. However, in DNN, the use of a smaller number of components does not reduce the training time and worsens the performance of these techniques.

Thanks to the high performance of the models achieved, they can be used for water classification, preventing possible consequences derived from ingesting water in non-optimal conditions.

In future work, the performance of other supervised and unsupervised classification techniques and other feature extraction methods will be studied to compare their performance against the proposal shown in this work. In addition, the development of an intelligent hybrid system capable of detecting the potability of water in different locations will be analyzed, allowing to standardize and offer a useful tool for the field of potabilization.

Acknowledgement. Míriam Timiraos's research was supported by the "Xunta de Galicia" through grants to industrial PhD (http://gain.xunta.gal/), under the "Doutoramento Industrial 2022" grant with reference: 04_IN606D_2022_ 2692965.

Antonio Díaz-Longueira's research was supported by the Xunta de Galicia (Regional Government of Galicia) through grants to Ph.D. (http://gain.xunta.gal), under the "Axudas á etapa predoutoral" grant with reference: $ED481A2023072$.

Grant PID2022-137152NB-I00 funded by MICIU/AEI/10.13039/501100-011033 and by ERDF/EU.

Xunta de Galicia. Grants for the consolidation and structuring of competitive research units, GPC (ED431B 2023/49)

CITIC, as a center accredited for excellence within the Galician University System and a member of the CIGUS Network, receives subsidies from the Department of Education, Science, Universities, and Vocational Training of the Xunta de Galicia. Additionally, it is co-financed by the EU through the FEDER Galicia 2021-27 operational program (Ref. ED431G 2023/01).

References

1. Dataset, October 2023. https://www.kaggle.com/datasets/xiaoxiaoliangzi/water-potability-prediction
2. Efficient KNN classification algorithm for big data. Neurocomputing **195**, 143–148 (2016). https://doi.org/10.1016/j.neucom.2015.08.112, learning for Medical Imaging

3. Arkaan Susila, M.T., Nadiyansyah Firdaus, I., Chuzairi, M.F., Wibawa, I.P.D., Kallista, M.: Denoising autoencoder - extreme learning machine for improving the classification performance integrated with IoT water quality systems. In: 2023 10th International Conference on Electrical Engineering, Computer Science and Informatics (EECSI), pp. 405–411 (2023). https://doi.org/10.1109/EECSI59885.2023.10295703
4. Bejani, M.M., Ghatee, M.: A systematic review on overfitting control in shallow and deep neural networks. Artif. Intell. Rev. **54**, 6391–6438 (2021)
5. Cortes, C., Vapnik, V.: Support-vector networks. Mach. Learn. **20**, 273–297 (1995)
6. Falkenmark, M., Widstrand, C.: Population and water resources: a delicate balance. Popul. Bull. **47**(3), 1–36 (1992)
7. F.R.S., K.P.: LIII. on lines and planes of closest fit to systems of points in space. Philos. Mag. Ser. 1 **2**, 559–572 (1901)
8. Guo, G., Wang, H., Bell, D., Bi, Y., Greer, K.: KNN model-based approach in classification. In: Meersman, R., Tari, Z., Schmidt, D.C. (eds.) OTM 2003. LNCS, vol. 2888, pp. 986–996. Springer, Heidelberg (2003). https://doi.org/10.1007/978-3-540-39964-3_62
9. Janani, S., Vydehi, S.: Chaos-enhanced bat algorithm for optimizing hyper parameters in water quality classification. In: 2023 International Conference on Self Sustainable Artificial Intelligence Systems (ICSSAS), pp. 743–750 (2023). https://doi.org/10.1109/ICSSAS57918.2023.10331752
10. Jinan, A., Siregar, M., Rolanda, V., Suryani, D.F., Muis, A.: Comparing neural networks, support vector machines, and Naïve Bayes Algorhythms for classifying banana types. J. Comput. Netw. Archit. High Perform. Comput. **6**(1), 98–107 (2024)
11. Khoirul Haq, M.I., Dwi Ramadhan, F., Az-Zahra, F., Kurniawati, L., Helen, A.: Classification of water portability using machine learning algorithms. In: 2021 International Conference on Artificial Intelligence and Big Data Analytics, pp. 1–5 (2021). https://doi.org/10.1109/ICAIBDA53487.2021.9689727
12. Preetham, M., Krishnan, M.: Improving water quality prediction and monitoring using machine learning algorithms. In: 2023 12th International Conference on System Modeling and Advancement in Research Trends (SMART), pp. 367–373 (2023). https://doi.org/10.1109/SMART59791.2023.10428490
13. Priskilla Angel Rani, J., Nivasini, R., Yesubai Rubavathi, C., Jona, P.: Machine learning based real time water quality monitoring system. In: 2023 Third International Conference on Artificial Intelligence and Smart Energy (ICAIS), pp. 1366–1370 (2023). https://doi.org/10.1109/ICAIS56108.2023.10073761
14. Samek, W., Montavon, G., Lapuschkin, S., Anders, C.J., Müller, K.R.: Explaining deep neural networks and beyond: a review of methods and applications. Proc. IEEE **109**(3), 247–278 (2021). https://doi.org/10.1109/JPROC.2021.3060483
15. Sulaiman, K., Ismail, L.H., Razi, M.A.M., Adnan, M.S., Ghazali, R.: Water quality classification using an artificial neural network (ANN). In: IOP Conference Series: Materials Science and Engineering, vol. 601, p. 012005. IOP Publishing (2019)
16. Uddin, M.G., Nash, S., Olbert, A.I.: A review of water quality index models and their use for assessing surface water quality. Ecol. Ind. **122**, 107218 (2021)
17. Xie, X.: Principal component analysis. Wiley interdisciplinary reviews (2019)
18. Yang, T., Ying, Y.: AUC maximization in the era of big data and AI: a survey. ACM Comput. Surv. **55**(8) (2022). https://doi.org/10.1145/3554729,

Energy Forecasting Using Intelligent Models

Rafael Arnay[1], Javier Hernández-Aceituno[1(✉)], José-Francisco Gómez-González[2], and Juan A. Méndez-Pérez[1]

[1] Departamento de Ingeniería Informática y de Sistemas, Universidad de La Laguna,Avda. Astrofísico Fco. Sánchez s/n, 38203 La Laguna, Canary Islands, Spain
{rarnayde,jhernaac,jamendez}@ull.es

[2] Departamento de Ingeniería Industrial, Universidad de La Laguna, Avda. Astrofísico Fco. Sánchez s/n, 38203 La Laguna, Canary Islands, Spain
jfcgomez@ull.es

Abstract. This paper deals with the problem of intelligent energy forecasting in hotels. The aim is to provide reliable predictions of hotel energy demand to increase the efficiency of the management systems responsible for the energy resources planning. The immediate benefits will be related to reduction of energy consumption, decrease in associated greenhouse gases emissions and improvement of the sustainability indexes of hotel activities. The prediction algorithm is based on the use of intelligent methods. The objective is to provide a 24 h-prediction of consumed energy with a look ahead of 24 h. LSTM and GRU networks were used in the algorithm for this task. The algorithm, that includes the main specific variables affecting the consumption, is endowed with some important capabilities that improve the existing methods. In particular, the proposed model is able to adapt online to changes while maintaining a balanced trade-off between accuracy and simplicity. The evaluation of the proposal was done in a luxury hotel in Canary Islands. The results obtained show promising results for a real-time implementation of the method in an energy management system.

Keywords: Energy Efficiency · Sustainability · Intelligent Methods · Energy Forecasting

1 Introduction

The tourism industry is one of the main engines of economic development in many countries representing a very important percentage of their gross domestic product (GDP) [5,9]. Tourism creates new opportunities in the receiving countries, reducing the poverty indices, especially in the tropical and subtropical areas, where most of the Small Island Developing States (SIDS, e.g. Fiji, Seychelles, Haiti, Jamaica, Cape Verde, etc.) are located [23]. However, although tourism activities carry out an increase in the wealth of the region, there is also an increase in energy demand and in CO_2 emissions, which lead to environment

F. Zayas-Gato et al. (Eds.): DCAI 2024, LNNS 1173, pp. 11–21, 2024.
https://doi.org/10.1007/978-3-031-73910-1_2

degradation and pollution [11,13,22]. Consequently, International organizations such as the United Nations (UN), the European Union (EU) and many other governments have appointed the importance of diminishing the huge impact of the touristic sector on the environment and the need for a greenhouse gas emissions (GHG) [8,11,13,14,21,22]. Among all the possible measures, two of them are energy efficiency and the use of renewable energy resources [24].

For this, energy management systems (EMS) have been proposed with the aim of guaranteeing the quality of the energy supply according to the demand, optimizing the energy resources and planning the hotel activities with energy reduction criteria [2]. All the planning and decision making to improve the energy efficiency of the hotel should be done in accordance with the ISO 50001 standard for energy management [25].

One of the main variables in any EMS is the forecast of energy demand. Time horizons for this forecast can range from days to weeks. Daily forecast is the basis for the operational decisions taken in the use of resources. To this end the availability of reliable and accurate daily forecasts is crucial for decision making. Two main elements are necessary for this: historical demand data and forecasting methods. Generalization of digitization and the advancement of Internet of Things have made possible the availability of reliable data in most of the facilities.

Concerning the forecasting methods, there are three main methodological strategies for energy forecast: statistical methods, machine learning methods, and hybrid methods [7]. Statistical methods are based on examination, explanation and association of data, while machine learning methods use intelligent algorithms to learn from data and build forecast models. There are also hybrid proposals that are based on the combination of these methods.

One of the most common statistical methods used is the Auto-Regressive Integrated Moving Average (ARIMA) [27], where various decomposition models are considered for forecasting. A variation of this proposal that considers exogenous inputs is the ARIMAX method [12]. Although these methods have proven to have satisfactory results for solving linear problems, their performance decreases for complex models and also the number of parameters to tune increases considerably.

Different applications using recurrent neural network (RNN) models can be found [16]. The main problem arising with RNN models in the area of time-series forecasting is that they suffer from vanishing gradient problem. Thus, when the sequence is long, there is a limit to how much information can be stored in memory which can make it difficult to learn. Different algorithms based on RNN were proposed to overcome these vanishing gradient problems such as LSTM and GRU [15].

The work presented in this paper is focused on the short-term (24 h ahead) prediction of energy in hotels. The study includes a comparative analysis of different energy forecast algorithms for hotels. The methods proposed for the forecast were based on LSTM and GRU, that offer a well-known high performance in time-series prediction. The model will be validated using real data of a luxury hotel in Tenerife (Canary Islands). The main contributions of the work compared to existing proposals are:

- The proposal of an intelligent model for energy forecast to be applied specifically in hotels. For this, we include in the model the main specific variables affecting energy consumption in hotels (temperature, occupancy, historic consumption data).
- The algorithm was endowed with adaptation capabilities so that it can be easily updated with time.
- The algorithm proposed achieves a relevant trade-off between accuracy and simplicity.

The paper is structured in 4 sections. In Sect. 2, the description of the main characteristics of the hotel where the model was validated is presented together with the description of the solution and methods presented in this work. Specifically, LSTM and GRU algorithms are described for the considered application. Section 3 will summarize the main results obtained with each method and will compare the performance of the different methods. A discussion of the results is also included in this section. Finally, the main conclusions contributions of the paper are summarized in Sect. 4.

2 Methods

This section presents the basis of the algorithm for the 24 h ahead energy demand prediction and the case study. As commented, the methodology will be based on intelligent methods. In particular, LSTM and GRU strategies are considered. The validation will be performed using real data and will be also compared with standard methods (ARIMA approximation).

Long Term Short Memory (LSTM) is one of the methods that provide the best performance for both energy consumption and generation [1,17]. LSTM networks are a type of RNN, which have better capabilities to handle time-series data. Basically, their memory allows them to remember the existing dependencies among the sequence elements and also to process long sequences. In [10] several deep learning models (including LSTM models) trained using simulated data were proposed to predict hourly-based multi energy vectors on multiple building types. The authors concluded that LSTM were more suitable for prediction than convolutional neural network models. In [18], a deep-learning-based interval forecasting model for building energy prediction was developed by combining fuzzy information granulation, attention mechanism, and long short-term memory network. The conclusion is that the forecasting model using attention-based LSTM offers more advantages to improve interval forecasting performance than the conventional LSTM. It is particularly relevant the contribution proposed in [20] where the authors present a model for energy prediction of different types of buildings based on long short-term memory and multi-kernel maximum mean difference (MK-MMD) domain adaptation. One of the contributions of this paper is that the model can be applied to different types of buildings lacking historical data. For this the concept of transfer learning was included in the proposal using a multisource transfer model.

A different algorithm that also provides satisfactory results for time-series forecasting is the Gated Recurrent Unit (GRU). This algorithm is a RNN gating mechanism that was introduced in 2014 by Chung et al. [4]. The main differences between GRU and LSTM is that it has a forget gate and fewer parameters than LSTM and it does not have an output gate. In [26], an ensemble method that uses GRU is proposed for electricity consumption in buildings. A preliminary study before the work presented here can be found in [32]. In this paper, the potential of these machine learning techniques was evaluated in a hotel. Although satisfactory results were obtained, some limitations related to adaptation and online training were necessary for an efficient implementation in practice [19].

2.1 Case Study

In order to demonstrate the potential of the proposed methods and their validity in a real situation, they have been applied to a hotel. The chosen case study is a luxury resort in the Tenerife (Canary Islands, Spain) consisting of a main building and a villa area. The total number of rooms is 461, of which 313 are in the main building and 148 are in the villa area. In addition to all the amenities of a luxury hotel (swimming pools, spas, gyms, several restaurants, air-conditioning systems, meeting rooms, showrooms, etc.), it has a fleet of 110 electric bogie vehicles, some of which are for the exclusive use of guests. Charging points are distributed throughout the hotel complex for all vehicles. Each vehicle has four 12 V 140 Ah batteries. All these services offered by the hotel require a total contracted power of 1.84 MW. The connection point supplied by the electricity distribution company is medium voltage. The distribution inside the hotel complex is medium voltage with transformation centres distributed throughout the hotel complex for low voltage consumption. The data available for this hotel are the energy consumed per hour, the daily occupancy rate of the hotel and the daily maximum and minimum temperature measured at a nearby meteorological station of the Spanish National Meteorological Agency.

2.2 Description of the Proposal

In the case at hand, given a temporal sequence of the last h hours of characteristics such as power demand, ambient temperature, occupancy, etc., we want to predict a sequence of 24 h of power consumed, with a look ahead of 24 h. That is, at 00:00 on a Monday, the model must predict the sequence of 24 h of power demand starting at 00:00 on Tuesday.

The input sequence will be of size $[h, f]$ where h is the number of hours in the sequence and f the number of features to use. We will call *lookback* the period of time covered by the input sequence to the model.

Since we want to predict the power demand of a 24-h interval, we have used two approaches:

- A model that predicts a sequence of 24 values of power demand simultaneously as depicted in Fig. 1a. In the results section, we refer to this approach as "Simultaneous".
- A model that predicts only one value of power demand. The model is used 24 times iteratively, at each iteration discarding the initial value of the input sequence and feeding it back with the value predicted in the previous iteration. This approach is depicted in Fig. 1b and we refer to this approach as "Incremental" in the results section.

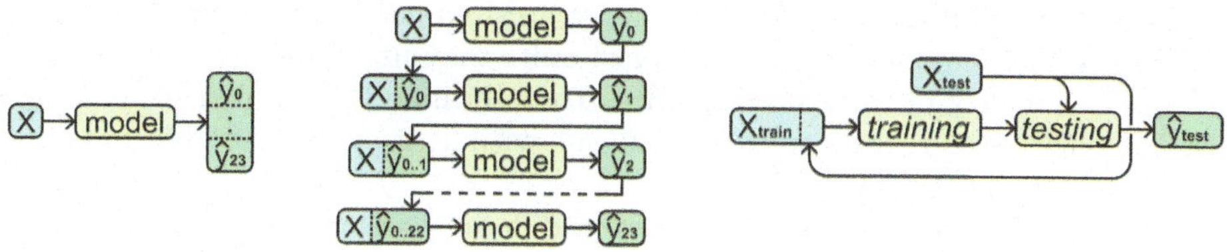

(a) Simultaneous (b) Incremental (c) Online training

Fig. 1. Prediction and training models.

In addition, as the objective of this work is to have a model that is easily trainable as new data is generated to improve its adaptability, experiments have been carried out comparing static and online training.

In static training the models are trained once with the train data and predictions are made with test data without retraining the models.

In online training, the models are initially trained with the original training data, but each time a new batch of 24 h of data from the test set is used to make predictions, that data becomes part of the training set and the models are retrained. The process is repeated until the test set is consumed. Figure 1c depicts this method.

2.3 Dataset

The dataset consist of a sequence of 365 days with the following data: hourly power consumption of the hotel in kW, hour of the day $[0-23]$, day of the week $[0-6]$ (0 being Monday), average occupation of the day, hourly temperature (°C), min temperature (°C) forecasted in the next 24 h and max temperature (°C) forecasted in the next 24 h. These data are expected to be readily available in real time, once the presented work is released to a live environment. The train and test splits consist of 281 and 84 days, respectively. Approximately one week out of four are selected for testing, evenly distributed throughout the year.

3 Results and Discussion

Different experiments were carried out on the real data containing energy demand information of the hotel. First step was to find the best parameterization for the LSTM and GRU algorithms (number of units and lookback days and the best features set). An analysis of the average errors for the different approaches was done. One important contribution of the paper is the performance analysis of online incremental versions of the algorithms. Besides, specific results to extract conclusions on the model performance depending on the specific time of day and each day of the week is also presented.

One of the parameters that has the greatest impact on the performance of LSTM and GRU models is the number of units they use. This number indicates the dimensionality of their hidden state. To determine the number of units and the lookback period, tests were carried out with 1, 2, 5, 10, 20, 100 units and with 5, 2, 1 lookback days.

The experiment involved testing seven different sets of features. The initial set included power consumption as a unique feature. The remaining sets were constructed by gradually introducing new features in the following order: hour of the day, day of the week, average occupation, average temperature, minimum temperature forecasted for the next day, and maximum temperature forecasted for the next day.

The configuration which produced the best results for both LSTM and GRU models, for a validation set that included 20% of the training sample, was defined by 1 lookback day and 20 units, with the following five features: power consumed, hour of the day, day of the week, average occupation, and average temperature.

Based on the best parameterization obtained for the LSTM and GRU models, 1 day of lookback and 20 units, variants of these models were trained with online and offline training and with simultaneous 24-h prediction and incremental prediction. As described in Sect. 2.2, in static training, the models undergo a single training session using the training data, and subsequent predictions are made using the test data without retraining the models. On the other hand, in online training, the models are initially trained using the original training data. However, whenever a new batch of 24 h of data from the test set is used to make predictions, it is incorporated into the training set. The models are then retrained using this updated training set. This process is repeated until all the data in the test set has been utilized. The "simultaneous" model is designed to forecast a sequence consisting of 24 power demand values simultaneously. In contrast, the "incremental" model is specifically built to forecast a single power demand value. This model is utilized iteratively, with 24 iterations in total. During each iteration, the initial value of the input sequence is discarded, and the model takes the previously predicted value as input in order to generate the next prediction. This iterative process continues until all 24 predictions are obtained.

Table 1 shows a comparison of the performance of the LSTM, GRU and ARIMAX models, including Mean Absolute Percentage Error (MAPE), Mean Absolute Error (MAE), Mean Squared Error (MSE) and Root Mean Squared Error (RMSE) [3,6,28]. As can be seen, the online-trained variants of LSTM and GRU perform better than their respective offline-trained versions. In addition, the versions with a simultaneous 24-h prediction also perform with less error than the incremental versions.

Table 1. Errors obtained by the models in the test set.

Model	MSE (kW^2)	MAE (kW)	RMSE (kW)	MAPE (%)
ARIMAX 21 days	23825.6	114.9	137.1	7.1
LSTM incr.	11146.5	80.8	104.7	5.0
LSTM incr. online	8909.5	71.7	94.3	4.4
LSTM simult.	6750.0	63.9	82.2	4.0
LSTM simult. online	6773.8	60.4	82.3	3.8
GRU incr.	11456.0	83.4	106.7	5.2
GRU incr. online	10912.4	80.3	104.4	5.0
GRU simult.	7080.0	65.9	84.1	4.1
GRU simult. online	6936.9	61.3	83.3	3.8

In addition to this analysis of the average error for the entire test set, we wanted to explore the behavior of the predictions of the trained models throughout the 24 h of each day, for each day of the week separately. To do this, the test data has been divided by day of the week, and the predictions have been averaged by hour. Figure 2 shows the comparison between the real demanded power and the predictions of the different models. As can be seen, the LSTM and GRU models fit better than the ARIMAX model to the actual power demand. The improvement is more evident between the hours 10 and 20 of each day, where the demand experiences a more contained increase compared to the slope of the demand between hours 5 and 10. ARIMAX models, based on sinusoidal functions, have more trouble fitting this change in demand. In general, it can be observed that LSTM and GRU models, in their different variations, are capable of fitting the demand prediction quite well on all days of the week. The "simultaneous" models adjust the demand prediction better than the "incremental" ones, which tend to overestimate. This is more evident on Sundays when the demand sharply drops compared to the previous day. However, this tendency of the "incremental" models to overestimate favors their fitting on Tuesdays when the actual demand increases compared to Mondays.

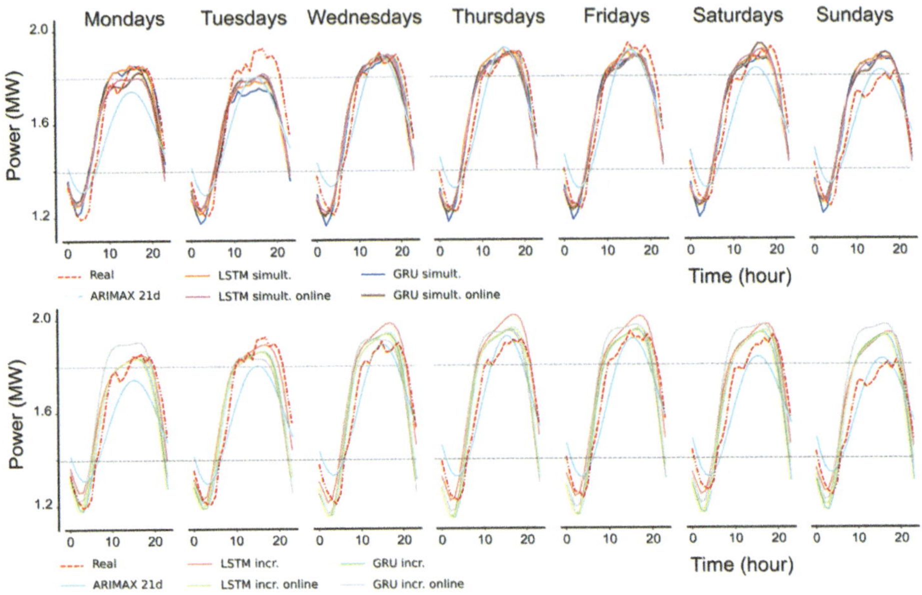

Fig. 2. Averaged power demand predictions and real power demand throughout the 24 h of each day of the week using simultaneous predictions (top) and incremental predictions (bottom).

4 Conclusions

The tourism industry, although beneficial for the economic development of many countries, has a significant impact on the environment, including an increase in energy consumption and greenhouse gas emissions. Therefore, it is essential to adopt measures to reduce the environmental impact of tourism activities, such as energy efficiency and the use of renewable energy resources. Within the tourism industry, accommodation is one of the main energy consumers, with hotels being the most significant. A strategy to reduce the environmental impact of hotels is the application of energy management systems where energy demand forecasting is a crucial input for these systems.

Two main features are important for an efficient performance of energy demand forecasting models in real applications, such as in the hotel sector. First, the availability of a mechanism for new relevant data update will improve the accuracy of the predictions. In this sense, it is necessary to study which are the most relevant variables and how do they influence the energy demand. This was studied in this work where we propose a set of five features as the most important to predict the evolution of the demand. Also, the most adequate size of the past time window was studied. A second important feature of the algorithm is the online adaptation as a valuable characteristics to deal with the changes in the consumption profiles, which in the future could be applied to easily produce different prediction models to different areas of the hotel, starting from a common generically trained template. The proposal presented here is endowed with these features as a key element to improve the performance in the predictions.

The study shows that, although statistical methods such as ARIMAX could be used for energy demand forecasting, their performance decreases for complex models such as the power system of a hotel. Consequently, machine learning methodologies, LSTM and GRU, have shown significant contributions to the development of prediction models for energy consumption and have greatly improved the accuracy of predictions. In addition, the improved performance of online-trained LSTM and GRU variants compared to their offline-trained counterparts is apparent. Furthermore, the 24-h simultaneous prediction versions exhibit lower error rates than the incremental versions.

Therefore, the online-training LSTM and GRU methodologies in the tourism industry, particularly in the accommodation sector, can be beneficial to improve energy efficiency and reduce environmental impacts. In the future, the authors intend is to apply a transference of learning, adapting the knowledge obtained from this model to other areas of the island, including those under different climatic conditions and hotels with different consumption patterns.

Acknowledgments. This work has been partially funded by the project Sustainable Atlantic Communities (SAtComm) EAPA_0019/2022 co-funded by the European Union through the Interreg Atlantic Area call. We acknowledge their support to our research.

Disclosure of Interests. The authors have no competing interests to declare that are relevant to the content of this article.

References

1. Abu-Salih, B., Wongthongtham, P., Morrison, G., Coutinho, K., Al-Okaily, M., Huneiti, A.: Short-term renewable energy consumption and generation forecasting: a case study of western Australia. Heliyon **8**(3), e09152 (2022)
2. Aman, S., Simmhan, Y., Prasanna, V.K.: Energy management systems: state of the art and emerging trends. IEEE Commun. Mag. **51**(1), 114–119 (2013)
3. Bickel, P.J., Doksum, K.A.: Mathematical Statistics: Basic Ideas and Selected Topics. Volumes I-II Package. Chapman and Hall/CRC, Boca Raton (2015)
4. Chung, J., Gulcehre, C., Cho, K., Bengio, Y.: Empirical evaluation of gated recurrent neural networks on sequence modeling. arXiv preprint arXiv:1412.3555 (2014)
5. World Travel & Tourism Council: World travel & tourism council - economic impact research (2024). https://wttc.org/research/economic-impact
6. De Myttenaere, A., Golden, B., Le Grand, B., Rossi, F.: Mean absolute percentage error for regression models. Neurocomputing **192**, 38–48 (2016)
7. Debnath, K.B., Mourshed, M.: Forecasting methods in energy planning models. Renew. Sustain. Energy Rev. **88**, 297–325 (2018). https://doi.org/10.1016/j.rser.2018.02.002
8. Dogan, E., Aslan, A.: Exploring the relationship among CO2 emissions, real GDP, energy consumption and tourism in the EU and candidate countries: evidence from panel models robust to heterogeneity and cross-sectional dependence. Renew. Sustain. Energy Rev. **77**, 239–245 (2017)

9. Filipiak, B.Z., Dylewski, M., Kalinowski, M.: Economic development trends in the EU tourism industry. Towards the digitalization process and sustainability. Qual. Quant. **57** (2023). https://doi.org/10.1007/s11135-020-01056-9
10. Gao, L., Liu, T., Cao, T., Hwang, Y., Radermacher, R.: Comparing deep learning models for multi energy vectors prediction on multiple types of building. Appl. Energy **301**, 117486 (2021)
11. Gössling, S.: Global environmental consequences of tourism. Glob. Environ. Chang. **12**(4), 283–302 (2002)
12. Hyndman, R.J., Athanasopoulos, G.: Forecasting: Principles and Practice. OTexts, Melbourne (2018)
13. Katircioglu, S.T.: International tourism, energy consumption, and environmental pollution: the case of turkey. Renew. Sustain. Energy Rev. **36**, 180–187 (2014)
14. Katircioglu, S.T., Feridun, M., Kilinc, C.: Estimating tourism-induced energy consumption and CO2 emissions: the case of Cyprus. Renew. Sustain. Energy Rev. **29**, 634–640 (2014)
15. Khalil, M., McGough, A.S., Pourmirza, Z., Pazhoohesh, M., Walker, S.: Machine learning, deep learning and statistical analysis for forecasting building energy consumption'a systematic review. Eng. Appl. Artif. Intell. **115**, 105287 (2022)
16. Kumar Tyagi, A., Abraham, A.: Recurrent Neural Networks. Concepts and Applications. CRC Press, Boca Raton (2022)
17. Li, J., Zhang, C., Sun, B.: Two-stage hybrid deep learning with strong adaptability for detailed day-ahead photovoltaic power forecasting. IEEE Trans. Sustain. Energy **14**(1), 193–205 (2022)
18. Li, Y., Tong, Z., Tong, S., Westerdahl, D.: A data-driven interval forecasting model for building energy prediction using attention-based LSTM and fuzzy information granulation. Sustain. Urban Areas **76**, 103481 (2022)
19. López, V., Casteleiro-Roca, J.L., Gato, F.Z., Mendez-Perez, J.A., Calvo-Rolle, J.L.: Intelligent model hotel energy demand forecasting by means of LSTM and GRU neural networks. In: Machado, J.M., et al. (eds.) DCAI 2022. LNCS, pp. 79–88, vol. 585. Springer, Cham (2023). https://doi.org/10.1007/978-3-031-23210-7_8
20. Lu, H., et al.: A multi-source transfer learning model based on LSTM and domain adaptation for building energy prediction. Int. J. Electr. Power Energy Syst. **149**, 109024 (2023)
21. Paramati, S.R., Shahbaz, M., Alam, M.S.: Does tourism degrade environmental quality? A comparative study of eastern and western European union. Transp. Res. Part D Transp. Environ. **50**, 1–13 (2017)
22. Pratt, L., Rivera, L., Bien, A.R.: Tourism in the Green Economy: Background Report. World Tourism Organization (UNWTO) and the United Nations Environment (2012)
23. Pratt, S.: The economic impact of tourism in SIDS. Ann. Tour. Res. **52** (2015). https://doi.org/10.1016/j.annals.2015.03.005
24. UN Environment Programme: Tourism. Investing in energy and resource efficiency (2011)
25. Rajić, M.N., Maksimović, R.M., Milosavljević, P.: Energy management model for sustainable development in hotels within WB6. Sustainability **14**(24), 16787 (2022)
26. Shan, S., Cao, B., Wu, Z.: Forecasting the short-term electricity consumption of building using a novel ensemble model. IEEE Access **7**, 88093–88106 (2019)

27. Smith, T.G., et al.: pmdarima: ARIMA estimators for Python (2017). http://www.alkaline-ml.com/pmdarima. Accessed 18 May 20230
28. Willmott, C.J., Matsuura, K.: Advantages of the mean absolute error (MAE) over the root mean square error (RMSE) in assessing average model performance. Climate Res. **30**(1), 79–82 (2005)

A Deep Learning-Based OCR System Implementation for Traceability Ensurement in a Metal Manufacturing Workshop

Paula Arcano-Bea[1], Míriam Timiraos[1], Pablo Fariñas[2], Francisco Zayas-Gato[1](✉), José Luis Calvo-Rolle[1], and Esteban Jove[1]

[1] Department of Industrial Engineering, CTC, CITIC, University of A Coruña, Ferrol, A Coruña, Spain
{paula.arcano,miriam.timiraos.diaz,f.zayas.gato,jlcalvo, esteban.jove}@udc.es

[2] Department of Naval and Industrial Engineering, University of A Coruña, Ferrol, Spain
pablo.farinas@udc.es

Abstract. The traceability of all elements within a metal manufacturing industry constitutes a critical challenge in the digitalization of industrial processes. In this context, the labeling of these elements, coupled with optical character recognition (OCR), plays a significant role. Therefore, in this paper, it is accomplished this challenge by applying three pre-trained deep learning OCR models such as Pytesseract, EasyOCR and KerasOCR to a dataset of images featuring labels with corresponding alphanumeric codes. The performance of these OCR engines has been evaluated using various types of input images, to which preprocessing techniques have been applied to enhance their quality and the legibility of the text contained within. The implementation of these models produced successful results, presenting a viable solution to improve the efficiency and accessibility of information retrieval processes within the industry.

Keywords: Optical character recognition · Tesseract · EasyOCR · KerasOCR · Image enhancement · Traceability · Metal manufacturing

1 Introduction

In recent years, the digitization of processes has increased exponentially due to several factors [3]. It has been shown that today, the competitiveness of institutions depends on digitalization, in addition to the ability to innovate [9]. Otherwise, the survival of organizations may not be viable in most cases. Digitization can contribute to more than just production improvement, such as reducing carbon emissions [1].

The naval industry, and in particular shipbuilding at shipyards, is completely aligned with this trend [18]. Digitalization is present from the concept and design

F. Zayas-Gato et al. (Eds.): DCAI 2024, LNNS 1173, pp. 22–31, 2024.
https://doi.org/10.1007/978-3-031-73910-1_3

phase to the most mature stages prior to the start of ship operation [5]. The competitiveness of European shipyards vis-à-vis Asian shipyards may depend, among other factors, on their modernization, as in the rest of the sectors [13]. European shipyards, and in particular the Spanish ones, take into account auxiliary companies as if they were part of the organization [11]. Therefore, the management and digitization tools must contemplate incorporating these companies that work for the shipyard.

Globally, digitization goes from the business aspects to the productive ones, and the tendency is toward integral management of the company [4]. Since the birth of quality management systems, traceability has been one of their pillars, giving the ability to trace all processes, from the acquisition of raw materials to the end of production [6]. Digitization is a crucial tool for traceability in the industry in general and, of course, for the shipbuilding industry [2].

In the course of production processes, parts must be unambiguously and correctly identified. This is especially important when an auxiliary company works for a shipyard and has to document and trace the processes. It should be emphasized that the used devices should, as far as possible, be of general purpose in order to avoid heavy dependence on specific equipment. Nowadays, for example, it is very common for anyone to have a cell phone that can take photographs. Then, it would simply be necessary to provide it with permissions so that it could upload photos to a management platform. If what is captured is a code, it is desirable that it be obtained automatically to avoid errors.

Taking all of the above into account, this work shows a real case of a company in the Spanish metal manufacturing industry that works for a shipyard. The company must trace its processes, and to do so, the manufactured parts are identified with a sticker that has a code. This code is captured with common smartphones. Its aim is to process the image and automatically obtain the alphanumeric string. Image preprocessing techniques are carried out, and then Optical Character Recognition (OCR) techniques are applied to obtain the character string.

The structure of this document is organized as follows: After this introductory section, the case study is described. The approach is then presented, and the different methods used for the implementation are described. Then, the results and their analysis are made. Finally, the conclusions and future works are presented.

2 Case of Study

In this research, real data from a Spanish company in the metal manufacturing sector, serving as an auxiliary entity for a shipyard, has been utilized. Ensuring traceability for every produced element is of significant importance to this company. This involves tracking not only the constituent materials, given the diverse base materials employed in the factory but also identifying the welder responsible for the welding process and determining the specific type of element among others. To enhance traceability, the implementation of optical character recognition (OCR) for the stickers affixed to these elements is proposed. This OCR process aims to effortlessly extract pertinent information from the stickers, contributing to a simplified and accessible data retrieval system. To this end, a dataset is obtained comprising multiple images featuring various types of elements, wherein the characters on these stickers are discernible to varying degrees of clarity. Two examples of images from the dataset are shown in Fig. 1.

Fig. 1. Images from the dataset

3 Approach

Acknowledging the information presented in the introduction and the case of study, our first approach involves constructing the dataset using images of the manufactured metal parts taken with a smartphone which will be visible the sticker that includes de alphanumeric code with specific information that metal piece. Afterwards, the application of image pre-processing techniques aimed at enhancing the images before initiating the character detection is proposed. The techniques to be employed include contrast and brightness adjustment to standardize illumination conditions across the entire dataset, as well as noise removal to improve image quality. Additionally, it will examine the scenario in which these techniques, along with binarization, are implemented to potentially eliminate background elements and emphasize the characters. Following the pre-processing phase, the character detection will be executed utilizing three of the most commonly used OCR engines, such as Tesseract, Keras OCR, and Easy-OCR. It will be conducted and evaluate the outcomes of those engines, considering various combinations of pre-processing techniques and OCR engines, with the goal of finding the optimal strategy for character detection within the context of the study. This process is described graphically in Fig. 2.

4 Methods

4.1 Pre-processing

Image preprocessing is an essential phase in the field of digital image processing, which involves the application of various techniques aimed at improving quality,

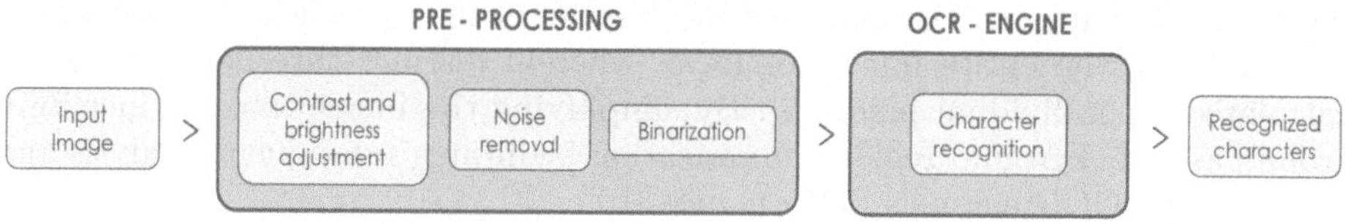

Fig. 2. Visual representation of the approach

sharpness, and legibility, among other characteristics, prior to their submission to more advanced analysis or processing [17]. In the field of optical character recognition, preprocessing becomes an essential measure to optimize the automatic interpretation of these characters by reducing possible distortions, improving contrast, and minimizing the noise present in the image that could affect the accuracy of the reading system. In this way, the aim is to ensure a more reliable and efficient data acquisition, contributing to greater accuracy in the identification and decoding of the alphanumeric content present in the images. Among the techniques frequently employed are contrast correction, noise reduction, brightness adjustment, binarization, and edge enhancement. This work combines a few of these techniques to improve the OCR results.

Contrast and Brightness Adjustment. The purpose of the contrast correction technique is to adjust the intensity disparity between the areas of higher and lower brightness in an image. Its purpose is to improve the discrimination and visibility of details, thus seeking to enhance the visual quality and facilitate the interpretation of the information contained in the image [16]. Alternatively, the brightness adjustment is used to improve visual quality and highlight relevant features in a data set. It consists of modifying the luminous intensity of pixels in an image by increasing or decreasing its value uniformly. Variations in illumination can affect the ability of algorithms to identify specific objects or patterns.

Noise Removal. It is a procedure that mitigates or reduces unwanted disturbances, commonly referred to as noise, that are present in an image. In order to implement these techniques, a range of algorithms is available, each one adaptable and adjustable to the distinctive characteristics inherent in individual cases. The goal is to enhance visual quality and ease subsequent analyses by diminishing undesirable interference in the visual information [17]. When dealing with images containing text, it's very important to apply noise reduction to achieve a polished and clearer image, enhancing the visibility of the text [7].

Binarization. This pre-processing technique involves the application of a threshold to transform a grayscale image into a binary representation, where pixels are categorized as either black or white based on their intensity values. The thresholding process for binarizing an image can be either general, where the threshold value is determined as the average of the highest and lowest intensity values of the entire image, or local, wherein distinct threshold values are calculated for individual pixels [7]. By simplifying the image into distinct foreground and background regions, binarization facilitates subsequent analysis and improves the OCR engine's performance [16].

4.2 OCR Engines

Tesseract. is an open source optical character recognition engine initially developed as a PhD research projectin 1994 in Hewlett-Packard Labs [15], now it is maintained and developed by Google [12]. It is one of the most accurate and popular OCR engines available and is used to convert text images into editable text in various formats.

Keras OCR. is a widely used optical character recognition tool that uses several machine learning algorithms for pattern recognition to identify the presence and arrangement of text in an image. This tool uses pre-trained models to identify the shape of alphanumeric characters in an image and then reconstruct them into regular text forms. As a result, it extracts words along with their corresponding bounding boxes [10]. Additionally, it allows end-to-end training for the development of custom OCR models.

EasyOCR. is an open source OCR tool that allows the user to extract text from images quick and easy. EasyOCR is popular among developers due to its ease of implementation and its ability to recognize various languages. With this tool, users can perform tasks such as extracting text from images, text translation and the integration of OCR functions into your own applications and systems. EasyOCR uses a ResNet backbone [14] and supports over 80 languages and major writing scripts such as Latin, Chinese, Arabic, Devanagari, Cyrillic and more [8].

5 Experiments and Results

5.1 Experiment Setup

Three techniques are used for image preprocessing. The initial technique employed was contrast and brightness adjustment. Given that the images within the dataset lacked a consistent lighting condition due to their varied capture environments, this resulted in disparate brightness levels. This technique was applied across all images to standardize the dataset under consistent conditions, compensating for discrepancies between them.

The second technique implemented was noise removal, aimed at mitigating the presence of unwanted small elements within the images. Given the industrial context of image capture, environmental particles, climatic conditions, and ongoing processes, among other factors, introduced noise. With this technique, the objective was to mitigate these influences as much as possible to enhance the overall image quality. For the noise removal within the images, a multi-step procedure was employed. Starting with a small kernel utilized for image dilation, followed by the application of the same kernel to erode the image, thereby eliminating minor details within the image. Subsequently, morphological operations and a median filter with a kernel size of three were applied to conclude the refinement process, ensuring a refined and noise-reduced image outcome.

The last pre-processing technique employed was binarization. In order to apply this technique, firstly is converted the image from RGB to gray scale. After the conversion it is used an adaptive threshold function, available in the OpenCV library, with a mean thresholding method which uses the arithmetic mean of the local pixel neighborhood to compute our threshold value; also include the parameter called "thresh binary" that uses a function to binarize the image. For the max value parameter, it is used 255 to separate the areas in black and white. An example of those images can be seen in Fig. 3.

Fig. 3. From left to right, images with two pre-processing techniques, three pre-processing techniques

In implementing the Tesseract OCR engine, our initial step involved the application of a custom configuration according to the specific characteristics of the dataset to optimize performance. While Tesseract offers an extensive array of parameters, it selected only those pertinent to our dataset requirements. In terms of page segmentation mode, the Sparse text setting was employed (parameter: psm = 11) to find as much text as possible in no particular order. The engine mode was set to utilize the Neural Nets LSTM (parameter: oem = 1) engine for enhanced accuracy. Additionally, a character whitelist was defined, restricting the engine from recognizing characters specified exclusively within the whitelist(parameter: whitelist = '0, 1, 2, 3, 4, 5, 6, 7, 8, 9, A, B, C, D, E, F, G, H, I, J, K, L, M, N, O, P, Q, R, S, T, U, V, Y). After applying this configuration, the text is extracted from the images in a string format.

In deploying the EasyOCR tool, the languages used by the engine are constrained to enhance the efficiency of the detection mechanism (parameter: Reader (["es", "en"])). Furthermore, the tool was given a few indications regarding the structural characteristics of the text (parameter: paragraph = False). Following these preparatory steps, the tool proceeds to engage in the process of text recognition, ensuring an optimized and accurate outcome.

To utilize the Keras OCR tool, the initial step is to download the pre-trained weights for both the detector and the recognizer, followed by the subsequent application of the recognizer. Additionally, Keras OCR offers the capability to fine-tune the detector and recognizer independently without having to do complete end-to-end training; this allows for a more tailored and efficient approach to model refinement, ensuring adaptability to specific requirements without the need for exhaustive training procedures.

5.2 Results and Analysis

The outcomes derived from the experimental setup described in the preceding paragraph are encapsulated in the data presented within Table 1, and visual representations through Fig. 4.

Table 1. Results of the OCR with the dataset images (%)

	Tesseract			Keras OCR			Easy OCR		
Pre-processed images	Matches	Errors	Missed	Matches	Errors	Missed	Matches	Errors	Missed
No pre-processing	79.47	2.65	17.88	83.44	7.28	9.27	81.46	5.29	13.25
Two techniques	86.08	4.64	9.28	79.13	6.09	14.78	82.78	3.31	13.91
Three techniques	82.47	10.31	7.22	76.82	6.62	16.56	89.40	0.00	10.59

In the previous table (Table 1) the results of the experiments are detailed in percentage format, distinguishing each OCR engine's performance including the attained matches, errors and missed characters. These results encompass diverse image inputs such as images without any pre-processing, images with only contrast and brightness adjustment and noise removal, and images undergoing analogous pre-processing alongside binarization. This categorization facilitates the assessment of OCR engine efficacy across varied image processing scenarios.

Figure 4 (top left) illustrates a graphical representation of the results for the dataset images with no pre-processing techniques. In this figure it is possible to see that among the OCR engines utilized, Keras OCR has the highest percent of accurate character detection. In the Fig. 4 (top right), there is a graphic representation of the results for the dataset images that were pre-processed with contrast and brightness adjustment and noise removal. The graph reveals that for this particular set of images, Tesseract has a better performance. Regarding the images that were pre-processed with contrast and brightness adjustment,

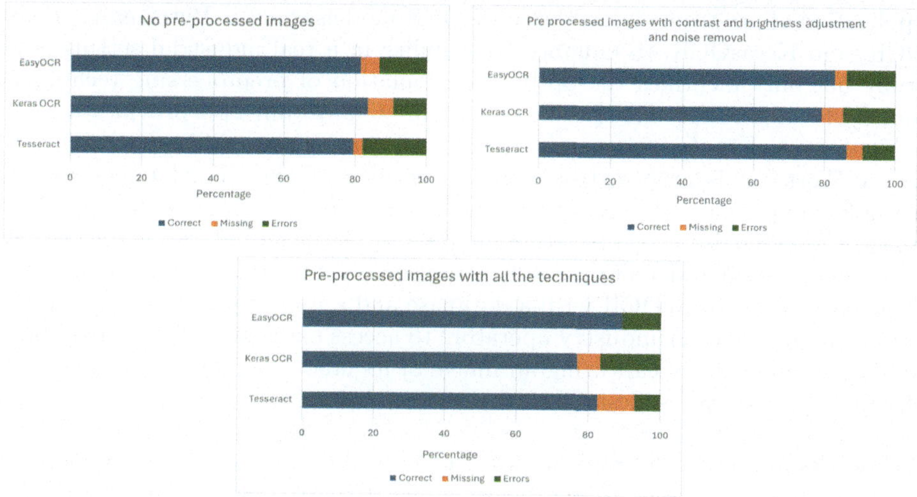

Fig. 4. Misses, errors and correct character detection. From top to bottom, left to right: With no pre-processed images, with pre-processed images (two techniques) and with pre-processed images (three techniques)

noise removal and binarization, the Fig. 4 (bottom) illustrates that the OCR engine that had the best performance under these pre-processing conditions was EasyOCR.

6 Conclusion and Future Works

This paper describes the results of optical character detection applied to our image dataset, employing various preprocessing methods and multiple OCR engines. In the specific context of our study, it is observed that EasyOCR achieves the detection of all characters, whether correctly or erroneously. For this particular scenario, achieving 100% correct character detection is not imperative, as it possesses a reference database that allows one to compare and identify inaccuracies in the read string. However, it's very important to recognize as many characters correctly as possible because elements within the same type tend to have a similar structure in the text on their tags, but each element has its own specific reference. For instance, in the image on the right in Fig. 3, the text reads "FALSOS PISO Y TECLES FGMS 1009E BLQ 211 ZONA 2". For this particular type of element, we may also encounter stickers with the text "FALSOS PISO Y TECLES FGMS 1012A BLQ 211 ZONA 2", as shown in the image on the right in Fig. 1. In this example, we observe that the difference between the two texts is only three characters so it is crucial to accurately recognize the maximum number of characters to effectively distinguish these two elements. This research contributes to the field of industrial digitalization, particularly within the metal manufacturing sector, by demonstrating the practical application of pre-trained deep learning OCR models, such as Pytesseract, EasyOCR, and KerasOCR, to enhance traceability in a real industrial setting. The study not only identifies the optimal combination of preprocessing techniques and OCR engines for this specific industrial scenario but also provides a replicable methodology for other industries facing similar traceability challenges. In future Works, it is proposed to increase the dataset with multiple captures of a single component. This approach involves reading the characters from these images, followed by a comparative analysis of the acquired character strings, to enhance the final detection result. Additionally, to increase traceability, it is suggested to integrate OCR with a database and a user application. This application would empower industry operators to access extensive information about a given component by capturing an image of its label, thereby fostering a more streamlined and accessible information retrieval process.

Acknowledgments. This work has been supported by Centro Mixto de Investigación UDC-Navantia (IN853C 2022/01), funded by GAIN (Xunta de Galicia) and ERDF Galicia 2021-2027.

Míriam Timiraos's research was supported by the "Xunta de Galicia" through grants to industrial PhD (http://gain.xunta.gal/), under the "Doutoramento Industrial 2022" grant with reference: 04_IN606D_2022_ 2692965.

CITIC, as a center accredited for excellence within the Galician University System and a member of the CIGUS Network, receives subsidies from the Department of Education, Science, Universities, and Vocational Training of the Xunta de Galicia. Additionally, it is co-financed by the EU through the FEDER Galicia 2021-27 operational program (Ref. ED431G 2023/01). Xunta de Galicia. Grants for the consolidation and structuring of competitive research units, GPC (ED431B 2023/49).

References

1. An, Q., Shi, Y.: Does enterprise digitization reduce carbon emissions? Evidence from China. Chin. J. Popul. Resour. Environ. **21**(4), 219–230 (2023). https://doi.org/10.1016/j.cjpre.2023.11.003. https://www.sciencedirect.com/science/article/pii/S2325426223000475
2. Beliatis, M.J., Jensen, K., Ellegaard, L., Aagaard, A., Presser, M.: Next generation industrial IoT digitalization for traceability in metal manufacturing industry: a case study of industry 4.0. Electronics **10**(5), 628 (2021)
3. Charfeddine, L., Umlai, M.: ICT sector, digitization and environmental sustainability: a systematic review of the literature from 2000 to 2022. Renew. Sustain. Energy Rev. **184**, 113482 (2023). https://doi.org/10.1016/j.rser.2023.113482. https://www.sciencedirect.com/science/article/pii/S1364032123003398
4. Cho, S., Lee, J.M., Woo, J.H.: Development of production planning system for shipbuilding using component-based development framework. Int. J. Naval Archit. Ocean Eng. **13**, 405–430 (2021). https://doi.org/10.1016/j.ijnaoe.2021.05.001. https://www.sciencedirect.com/science/article/pii/S2092678221000261
5. Diaz, R., Smith, K., Bertagna, S., Bucci, V.: Digital transformation, applications, and vulnerabilities in maritime and shipbuilding ecosystems. Procedia Comput. Sci. **217**, 1396–1405 (2023). https://doi.org/10.1016/j.procs.2022.12.338. https://www.sciencedirect.com/science/article/pii/S1877050922024231, 4th International Conference on Industry 4.0 and Smart Manufacturing

6. Grauer, Z.: Systems and traceability of quality systems. American Laboratory, p. 15 (2003)
7. Ignasius, A., Chandra, J.C., Oscadinata, R., Suhartono, D.: Image pre-processing effect on OCR's performance for image conversion to braille unicode. Procedia Comput. Sci. **227**, 922–931 (2023). https://doi.org/10.1016/j.procs.2023.10.599, 8th International Conference on Computer Science and Computational Intelligence (ICCSCI 2023)
8. Li, C., et al.: PyCinemetrics: computational film studies tool based on deep learning and PySide2. SoftwareX **26**, 101686 (2024). https://doi.org/10.1016/j.softx.2024.101686
9. Marti, L., Puertas, R.: Analysis of European competitiveness based on its innovative capacity and digitalization level. Technol. Soc. **72**, 102206 (2023). https://doi.org/10.1016/j.techsoc.2023.102206. https://www.sciencedirect.com/science/article/pii/S0160791X23000118
10. Martínez-Rojas, A., Jiménez-Ramírez, A., Enríquez, J., Reijers, H.: A screenshot-based task mining framework for disclosing the drivers behind variable human actions. Inf. Syst. **121**, 102340 (2024). https://doi.org/10.1016/j.is.2023.102340
11. Mickeviciene, R.: Global competition in shipbuilding: trends and challenges for Europe. The Economic Geography of Globalization, pp. 201–222 (2011)
12. Patel, C., Patel, A., Patel, D.: Optical character recognition by open source OCR tool tesseract: a case study. Int. J. Comput. Appl. **55**, 50–56 (2012). https://doi.org/10.5120/8794-2784
13. Salonen, A., Gabrielsson, M., Al-Obaidi, Z.: Systems sales as a competitive response to the Asian challenge: case of a global ship power supplier. Ind. Mark. Manage. **35**(6), 740–750 (2006). https://doi.org/10.1016/j.indmarman.2005.06.008. https://www.sciencedirect.com/science/article/pii/S001985010500101X
14. Schönfelder, P., Stebel, F., Andreou, N., König, M.: Deep learning-based text detection and recognition on architectural floor plans. Autom. Constr. **157**, 105156 (2024). https://doi.org/10.1016/j.autcon.2023.105156
15. Smith, R.: An overview of the tesseract OCR engine, vol. 2, pp. 629–633 (2007). https://doi.org/10.1109/ICDAR.2007.4376991
16. Sporici, D., Cuşnir, E., Boiangiu, C.A.: Improving the accuracy of tesseract 4.0 OCR engine using convolution-based preprocessing. Symmetry **12**(5), 715 (2020)
17. Sugiyono, A.Y., Adrio, K., Tanuwijaya, K., Suryaningrum, K.M.: Extracting information from vehicle registration plate using OCR tesseract. Procedia Comput. Sci. **227**, 932–938 (2023). https://doi.org/10.1016/j.procs.2023.10.600, 8th International Conference on Computer Science and Computational Intelligence (ICCSCI 2023)
18. Yi, Z., et al.: Intelligent initial model and case design analysis of smart factory for shipyard in china. Eng. Appl. Artif. Intell. **123**, 106426 (2023). https://doi.org/10.1016/j.engappai.2023.106426. https://www.sciencedirect.com/science/article/pii/S0952197623006103

Dimensional Reduction Techniques for the Characterization of Behavioral Patterns in Dairy Cows

Álvaro Michelena[1(✉)], Antonio Díaz-Longueira[1], Paula Arcano-Bea[1], Héctor Quintián[1], Óscar Fontenla-Romero[2], and José Luis Calvo-Rolle[1]

[1] University of A Coruña, CTC, CITIC, Department of Industrial Engineering, Avda. 19 de febrero s/n, 15405 Ferrol, A Coruña, Spain
{alvaro.michelena,a.diazl,paula.arcano,hector.quintian,jlcalvo}@udc.es

[2] University of A Coruña, LIDIA, CITIC, Faculty of Computer Science, Campus de Elviña, s/n, 15071 A Coruña, Spain
oscar.fontenla@udc.es

Abstract. The current global population growth has intensified the demand for feed, especially dairy and meat products. In response to this growing need, precision livestock farming has become essential to manage animals efficiently and sustainably. Within this new approach, individual monitoring of each animal is essential. To this end, activity monitoring collars have been postulated as one of the most valuable tools for this purpose. With this in mind, three dimensional reduction methods (ICA, Isomap and t-SNE) have been analyzed in this study to identify patterns in the daily behavior of dairy cows through data collected by these collars. T-SNE was found to be especially effective in distinguishing individual or small group behaviors of animals, which is crucial for improving herd management and early detection of health or behavioral problems.

Keywords: Precision Livestock Farming · t-SNE · ICA · Isomap · Individual Characterization · Pattern Recognition

1 Introduction

The accelerated growth of the world's population presents inevitable challenges in agricultural production, with livestock being a key component in the global food chain [4]. The need to satisfy the growing demand for dairy and meat products underlines the strategic importance of the livestock sector in the current context. However, this growth scenario raises critical concerns about efficiently addressing production and optimizing the quality and quantity of livestock-derived products [19].

For all these reasons, Precision Livestock Farming (PLF) emerges as an essential response to improve production efficiency and ensure long-term sustainability. Integrating advanced technologies in livestock management is fundamental to

F. Zayas-Gato et al. (Eds.): DCAI 2024, LNNS 1173, pp. 32–41, 2024.
https://doi.org/10.1007/978-3-031-73910-1_4

achieving these goals [2]. The digitization of the livestock industry thus becomes imperative, opening up a horizon of possibilities for optimizing animal performance and welfare. The strategic role of PLF lies in improving productivity and mitigating the environmental impacts associated with the expansion of the industry.

In recent years, multiple devices and digital tools have been developed for their implementation in cattle management, such as smart scales, radio frequency identification tags, hyperspectral cameras, and GPS chips, among others [8]. However, most popular and widely used devices today is the smart monitoring collar capable of continuously monitoring the animal's behavior [9]. The capability to collect accurate and real-time data through these collars represents a significant advancement in livestock monitoring [13]. These devices, equipped with sophisticated sensors, enable detailed capture of daily behaviors, providing a comprehensive view of animal activities [16].

At present, many researchers have focused on applying and implementing Machine Learning (ML) and Artificial Intelligence (AI) methods in the livestock industry to analyze data collected through digital devices and tools [3]. The introduction of these advanced technologies in livestock has the potential to increase efficiency, improve animal welfare, and optimize resources. However, despite these promising developments, the practical implementation of ML and AI in animal husbandry often fails to reach its full potential for several reasons. First, the highly variable and dynamic nature of livestock environments poses unique challenges. Farms are complex systems with numerous interconnected variables, from environmental conditions to animal health and behavior. Also, animals themselves are living beings with individual and variable behaviors, making it challenging to create predictive and generalizable models [5].

All this means that to improve the performance of intelligent techniques in the field of animal husbandry, it is necessary to develop personalized technological solutions that take into account the peculiarities and specific characteristics of each farm, each herd, and each cow [14]. Therefore, to successfully implement intelligent techniques, it is essential to characterize cows, herds, and farms in a specific way.

Therefore, this study focuses on the specific analysis of how dimensional reduction techniques, when applied to data generated by monitoring collars, can provide specific patterns in daily behavior and, consequently, the performance of dairy cows, thus contributing to the evolution toward more innovative and sustainable farming practices.

In summary, the main goal of this research is to explore the effectiveness of these methods for identifying and characterizing individual cows based on their unique behavioral patterns. This detailed characterization is crucial, as it allows individual differentiation within the herd and the identification of behavioral patterns that may indicate various physiological or health states.

Beyond behavioral characterization, this study lays the groundwork for future research that could apply artificial intelligence and machine learning in predicting and early detecting disease, identifying estrus, calving, and other abnormal events. The ability to anticipate and respond quickly to these events can transform livestock management.

The present document is structured as follows: in Sect. 2 the case of study is presented. Then, Sect. 3 provides a detailed description of the machine learning methods used. The experiments and their results are shown in Sect. 4. Finally, Sect. 5 exposes the conclusions and future work.

2 Case Study

2.1 RUMI Smart Collar

Developed by the Galician company Innogando [1], the RUMI ® collar is a GPS monitoring device for livestock that integrates advanced technology in a design focused on animal welfare. This device stands out for its lightness and small size, avoiding discomfort for the livestock. Despite its compact size, RUMI is equipped with powerful sensors capable of detecting five different behaviours (rest, walk, ruminate, graze, and eat/play) and the number of steps, providing crucial data for efficient herd management.

An outstanding feature of RUMI is its robust plastic casing, which encapsulates the device and makes it highly resistant and waterproof. This protection guarantees optimum collar performance in different environmental and climatic conditions, thus ensuring its durability and reliability. In addition, the device includes a solar panel that allows continuous recharging of the battery, ensuring uninterrupted operation and reducing the need for manual battery recharges.

To ensure accurate system performance, it is essential to place the collar correctly on the cattle: it must be at mid-neck height and on the left side as shown in Fig. 1. This specific position optimizes data collection and maintains the animal's comfort. Overall, Innogando's RUMI collar presents itself as a comprehensive and meticulously designed solution for effective livestock monitoring, providing farmers with detailed control and significantly improving the management of their animals.

Fig. 1. Activity collar designed by Innogando [1]

2.2 Dataset

Within the framework of this research, information has been collected from a group composed of 14 cows belonging to an intensive dairy farm located in Asturias, Spain. Each of these cows is identified by a unique code number. The data were recorded from 1 July to 8 November 2023. All cows monitored belong to the Holstein/Friesian breed and have ages ranging from two and a half to seven years.

It is essential to note that all cows included in this study come from the same herd and have undergone the same procedures characteristic of an intensive dairy farm, such as milking sessions, among others. However, during the period of activity registration, several cows showed atypical behaviour, such as illness or estrus days. The dataset used in this study covers all the minutes each cow carried out a specific activity (rest, walk, ruminate, graze, and eat/play) and the number of steps taken in a day, so the number of features is six.

3 Intelligent Techniques Implemented

In this research work, three different dimensional reduction techniques have been implemented, and their results will be compared. Each of these techniques will be described below. The selection of these three methods is due to their successful performance in previous work and in a multitude of applications and fields of study [12].

Independent Component Analysis (ICA). Independent Component Analysis, also known as ICA, is a dimensionality reduction technique used to explore and analyze multichannel data. ICA aims to model data as a linear combination of independent non-Gaussian sources [15]. It is a multivariate approach that performs a linear transformation to obtain uncorrelated and independent components [18]. The primary objective of ICA is to find a linear assignment of a source vector so that each estimate component is as independent as possible.

In practical applications, ICA is utilized in data analysis or data compression. This method is an iterative process that extracts independent components from linear transformations of the original dataset. Multiple algorithms have been proposed to estimate independent components based on different objective functions.

In summary, ICA is a powerful and widely used method for decomposing observed data into statistically independent components. The use of ICA allows for a more accurate and efficient analysis of complex multichannel data.

Isomap. Isomap is a nonlinear dimensionality reduction technique that preserves the data neighbourhood structure [17]. Unlike Principal Component Analysis (PCA), which focuses on the variance of the data, Isomap seeks to preserve the structure of the data in a lower dimensional space. Isomap is based on the idea that data close in the high-dimensional space should also be close in the low-dimensional space [7].

The most critical step in the Isomap method is the construction of the Geodesic distance matrix. This matrix is constructed using a nearest neighbour technique that searches for the k nearest neighbours of each data point. Dijkstra's algorithm is then used to find the geodesic distance between each pair of data points. This distance between two data points is defined as the length of the shortest path connecting them in the neighbourhood network [6].

Once the matrix has been constructed, the Multidimensional Scaling dimensionality reduction (MDS) technique projects the data into a lower dimensional space. MDS seeks a representation in a lower dimensional space that preserves the geodesic distances between the samples.

T-Distributed Stochastic Neighbor Embedding (t-SNE). T-distributed Stochastic Neighbor Embedding, known as t-SNE, is a statistical and non-linear technique developed in 2008 by Laurens van der Maaten and Geoffrey Hinton [11]. It is widely used for analyzing and visualizing high-dimensional data in a lower-dimensional space.

The t-SNE approach involves transforming high-dimensional data points into joint probabilities, which are then used to minimize the Kullback-Leibler divergence (KLd) [10]. This ultimately results in low-dimensional embeddings that capture the relationships between the original data points.

Initially, the algorithm calculates joint probabilities for data points, representing their similarities. Subsequently, t-SNE generates a representation of data points in lower dimensions based on the probability distribution, iterating until the lowest KLd is achieved.

It is important to note that t-SNE utilizes a non-convex cost function, which means that diverse outcomes can be obtained with each iteration. However, a significant advantage is that, unlike other methods such as PCA, t-SNE performs exceptionally well in preserving low pairwise separation, leading to superior visualization due to its non-linear approach.

Finally, it is important to highlight that for its correct performance, the t-SNE algorithm requires the adjustment of a hyperparameter known as "perplexity". This parameter controls the selection of neighbors used in the local optimization.

4 Experiment Setup and Results

4.1 Experiment Configuration

For the execution of the experiments, a Python algorithm was developed using various libraries widely used in data analysis, such as Scikit-learn (sklearn), Pandas and Numpy, among others.

The first step in the development of the experiments was data preprocessing. During this phase, daily records with missing values and spurious values were eliminated to ensure the integrity and quality of the information analyzed. Subsequently, we normalized the dataset using the z-score method, standardizing the values to a mean of 0 and a standard deviation of 1.

With the data properly preprocessed and normalized, the dimensional reduction techniques described above were applied. For each of these techniques, experiments were performed with different hyperparameter values, always reducing the dimensionality to two components for a simple and visual interpretation of the behavioral patterns.

In this sense for ICA, three different whitening techniques were tested: "arbitrary variance", "unit variance", and a non-whitening configuration, exploring how each affected sample separation. For Isomap, the influence of varying the number of nearest neighbors with values of 10, 25, 50, 50, 75, 100, 100, 200 and 500 was evaluated, looking for the optimal balance between locality and globality in data structure. Finally with T-SNE, the perplexity parameter values explored were 10, 20, 30, 40, 50, 75 and 100, to determine how each configuration affected the data clustering.

Each technique and configuration was analyzed independently, with the aim of selecting those configurations that best discerned and separated the samples corresponding to each cow, thus minimizing the scatter within each animal's observations. This selection and analysis of configurations was key to identify the most effective techniques and parameters to characterize the daily cow behavior, allowing the detection of patterns and significant differences between animals in the same herd.

4.2 Results

This subsection presents the results obtained for the best configurations of each technique, following the criteria discussed above. The evaluation of these techniques is represented in Figs. 2, 3 and 4, corresponding to the results of ICA, Isomap and t-SNE, respectively.

The implementation of ICA, Fig. 2, has shown a limitation in the clear differentiation of behavioral patterns among the cows under study. Most of the behavioral data overlap in a common region of the two-dimensional plane, with the exception of the cows identified by codes 6777 and 7838. These two animals show distinctive patterns that separate them slightly from the general grouping, indicating unique or differentiated behaviors compared to the rest of the herd.

On the other hand, the results obtained applying Isomap, Fig. 3, reflects a similar trend to that observed with ICA, where the majority of cows do not clearly differentiate themselves in the reduced space. Cows 6777 and 7838 again stand out from the herd, suggesting that their ability to identify patterns and separate individual cows within this dataset remains limited.

Finally, Fig. 4 shows the results of implementing t-SNE. In this case, a more detailed and differentiated view of cow behavior has been obtained. This method has effectively separated the behavioral patterns of a significant group of animals, demonstrating considerable variability in daily behavior even within a homogeneous herd. Specifically, cows with codes 96, 1093, 2353, 2359, 3448, 6777 and

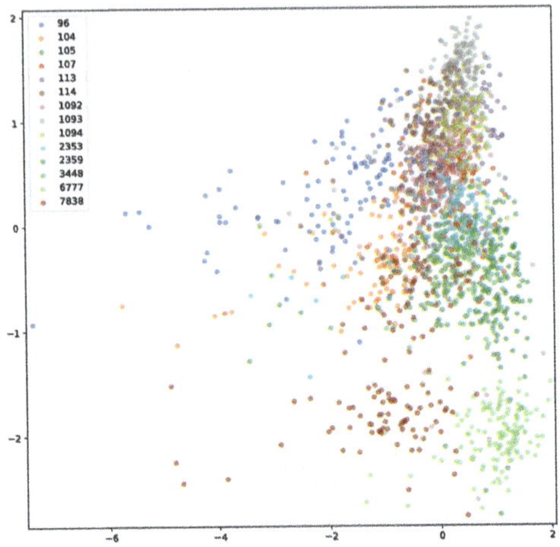

Fig. 2. ICA representation in 2D space

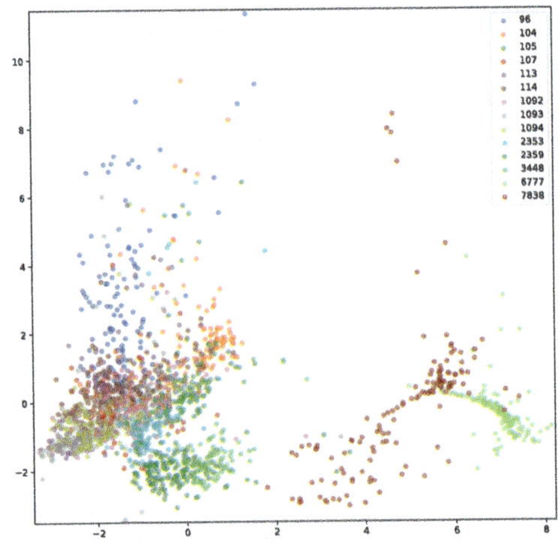

Fig. 3. ISOMAP representation in 2D space

7838 present distinctive behaviors that clearly differentiate them from the average observed in the rest of the herd, whose data tend to cluster in the same plane region, indicating similar behavioral patterns.

In addition, in Fig. 4 a particularity has been observed near the region associated with cow 1093, where data from other animlas (96, 1092, or 1094) deviate from their usual behavior patterns, which could indicate the presence of anomalous days due to specific events such as estrus or disease.

The results of this research highlight the existence of significant differences in the daily behavior of dairy cows, despite sharing the same environment and routines. The ability of t-SNE to differentiate between these patterns enhances

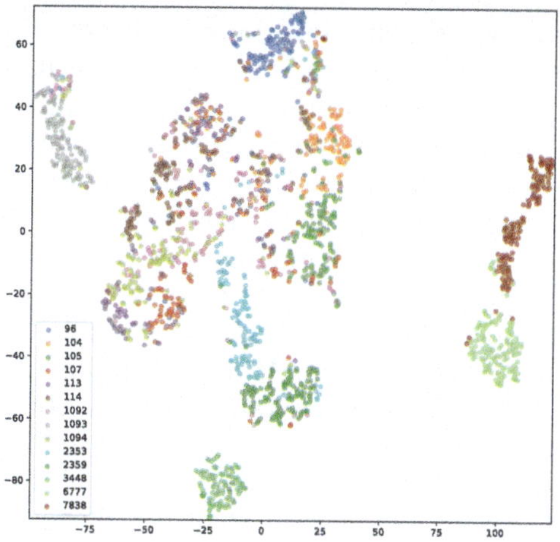

Fig. 4. T-SNE representation in 2D space

its usefulness in characterizing and understanding animal behavior. This characterization step is essential for future research seeking to detect abnormalities, identify reproductive states, or monitor herd health through behavioral analysis. The results point to the need to consider the individuality of animal behavior as a critical factor in the design and implementation of monitoring and management systems on dairy farms.

5 Conclusions and Future Work

In this research work, the characterization of daily individual behavior of dairy cows from the same herd has been addressed, using dimensional reduction techniques. For data collection, a commercial smart activity collar was used, capable of detecting and identifying five specific activities in addition to the number of steps. In particular, the behavior of a total of 14 dairy cows from the same herd in an intensive farm was recorded during four months.

Three different dimensional reduction techniques (ICA, ISOMAP and t-SNE) were applied to the daily data of each of the activities performed by the cow combined with the number of steps for their analysis and visualization in a two-dimensional space. The results obtained showed that the ICA and ISOMAP techniques did not achieve a clear differentiation between animals or groups of them. However, the application of t-SNE to daily activity and step data revealed significant differences in cow behavior, even when they all shared identical environmental conditions and daily routines. Therefore, this research has demonstrated how the use of t-SNE is effective in characterizing individual animal behavior despite the homogeneity of the environment and care received.

The identification and characterization of each animal is essential, since it establishes a solid base for future research oriented to the detection of anomalies, estrus, calving, among others, based on individual behavior. Before advancing towards the detection of such events, it is essential to accurately characterize the normal behavior of each cow, thus improving the accuracy of predictive models.

Future work will focus on examining the performance of these dimensional reduction techniques on a larger dataset. This will include the analysis of behaviors in different herds and a larger number of cows, extending to other farms, both dairy and beef, located in different geographic regions. In addition, the efficacy of these techniques will be evaluated in extensive farming environments, where the living conditions and cow behaviors may differ significantly from the intensive environments studied so far. Also, the application of clustering techniques to assess the degree of similarity in the behavior of different cows will be studied.

Acknowledgements. Álvaro Michelenas research was supported by the Spanish Ministry of Universities (https://www.universidades.gob.es/), under the "Formación de Profesorado Universitario" grant with reference: FPU21/00932.

CITIC, as a center accredited for excellence within the Galician University System and a member of the CIGUS Network, receives subsidies from the Department of Education, Science, Universities, and Vocational Training of the Xunta de Galicia. Additionally, it is co-financed by the EU through the FEDER Galicia 2021-27 operational program (Ref. ED431G 2023/01).

This work has been supported by Centro Mixto de Investigación UDC-NAVANTIA (IN853C 2022/01), funded by GAIN (Xunta de Galicia) and ERDF Galicia 2021-2027.

Antonio Daz-Longueira's research was supported by the Xunta de Galicia (Regional Government of Galicia) through grants to Ph.D. (http://gain.xunta.gal), under the "Axudas á etapa predoutoral" grant with reference: ED481A-2023-072.

Xunta de Galicia. Grants for the consolidation and structuring of competitive research units, GPC (ED431B 2023/49).

References

1. Innogando. (2023). https://innogando.com/. Accessed 19 Dec 2023
2. Berckmans, D.: General introduction to precision livestock farming. Anim. Front. **7**(1), 6–11 (2017)
3. Chaudhry, A.A., Mumtaz, R., Zaidi, S.M.H., Tahir, M.A., School, S.H.M.: Internet of Things (IoT) and machine learning (ML) enabled livestock monitoring. In: 2020 IEEE 17th International Conference on Smart Communities: Improving Quality of Life Using ICT, IoT and AI (HONET), pp. 151–155. IEEE (2020)
4. Fukase, E., Martin, W.: Economic growth, convergence, and world food demand and supply. World Dev. **132**, 104954 (2020)
5. Garcia, R., Aguilar, J., Toro, M., Pinto, A., Rodriguez, P.: A systematic literature review on the use of machine learning in precision livestock farming. Comput. Electron. Agric. **179**, 105826 (2020)
6. Geng, X., Zhan, D.C., Zhou, Z.H.: Supervised nonlinear dimensionality reduction for visualization and classification. IEEE Trans. Syst. Man Cybernet. Part B (Cybernetics) **35**(6), 1098–1107 (2005)
7. Ghojogh, B., Ghodsi, A., Karray, F., Crowley, M.: Multidimensional scaling, Sammon mapping, and Isomap: tutorial and survey. arXiv preprint arXiv:2009.08136 (2020)

8. Halachmi, I., Guarino, M., Bewley, J., Pastell, M.: Smart animal agriculture: application of real-time sensors to improve animal well-being and production. Annu. Rev. Animal Biosci. **7**, 403–425 (2019)
9. Handcock, R.N., et al.: Monitoring animal behaviour and environmental interactions using wireless sensor networks, GPS collars and satellite remote sensing. Sensors **9**(05), 3586–3603 (2009)
10. Ji, S., Zhang, Z., Ying, S., Wang, L., Zhao, X., Gao, Y.: Kullback-Leibler divergence metric learning. IEEE Trans. Cybernet. **52**(4), 2047–2058 (2020)
11. Van der Maaten, L., Hinton, G.: Visualizing data using t-sne. J. Mach. Learn. Res. **9**(11) (2008)
12. Michelena, Á., et al.: Beta Hebbian learning for intrusion detection in networks with MQTT protocols for IoT devices. Logic J. IGPL **32**(2), 352–365 (2024)
13. Neethirajan, S.: The role of sensors, big data and machine learning in modern animal farming. Sens. Bio-Sens. Res. **29**, 100367 (2020). https://doi.org/10.1016/j.sbsr.2020.100367
14. Riaboff, L., Shalloo, L., Smeaton, A.F., Couvreur, S., Madouasse, A., Keane, M.T.: Predicting livestock behaviour using accelerometers: a systematic review of processing techniques for ruminant behaviour prediction from raw accelerometer data. Comput. Electron. Agric. **192**, 106610 (2022)
15. Stone, J.V.: Independent component analysis: a tutorial introduction (2004)
16. Stygar, A.H., et al.: A systematic review on commercially available and validated sensor technologies for welfare assessment of dairy cattle. Front. Veterin. Sci. **8**, 634338 (2021)
17. Tenenbaum, J.B., Silva, V.D., Langford, J.C.: A global geometric framework for nonlinear dimensionality reduction. Science **290**(5500), 2319–2323 (2000)
18. Tharwat, A.: Independent component analysis: an introduction. Appl. Comput. Inform. **17**(2), 222–249 (2021)
19. Thornton, P.K.: Livestock production: recent trends, future prospects. Philos. Trans. Roy. Soc. B: Biol. Sci. **365**(1554), 2853–2867 (2010)

Geothermal Heat Exchanger's Temperature Input Sensor Prediction Based on Deep Learning Modelling Technique

Pedro Oliveira[1], Paula Arcano-Bea[2], Francisco S. Marcondes[1], José Luis Calvo-Rolle[2](✉), Paulo Novais[1], and Esteban Jove[1]

[1] ALGORITMI/LASI Centre, University of Minho, Braga, Portugal
{pedro.jose.oliveira,francisco.marcondes}@algoritmi.uminho.pt, pjon@di.uminho.pt, esteban.jove@udc.es

[2] University of A Coruña, Department of Industrial Engineering, CTC, CITIC, Ferrol, A Coruña, Spain
{paula.arcano,jlcalvo}@udc.es

Abstract. The goal of this study was to test many models in order to check which one better predicted the heat exchanger of a bioclimatic building. A number of regression models, pre-processing methods, and data analyses are compared in the study to forecast the input collector temperature of the heat pump. Specifically, three different techniques have been considered in this research work: Multilayer Perceptrons, Long Short Term Memory networks and Convolutional Neural Networks. Satisfactory results have been obtained in all cases for predicting the temperature 24 h in advance, implementing an useful tool for enhance energy management.

Keywords: Convolutional Neural Network · Forecasting · Heat exchanger · Long Short-Term Memory · Multi Layer Perceptron

1 Introduction

Over the past decades, the world has faced swift shifts regarding global economic volatility, heightened competitiveness, and climate change [4]. Given these conditions, the surge in raw material costs and a significant reliance on fossil fuels, coupled with the resultant environmental harm, have prompted an increased focus on alternative energy sources [7]. The availability of renewable energy is boundless, and its exploitation is comparatively clean. Hence, in recent years, various governments have enacted policies to promote the adoption of such energy forms [9].

Solar and wind energy are currently the most commonly utilized alternative energy sources, leading to significant technological advancements in these areas. However, the renewable energy sector has seen remarkable progress in

F. Zayas-Gato et al. (Eds.): DCAI 2024, LNNS 1173, pp. 42–51, 2024.
https://doi.org/10.1007/978-3-031-73910-1_5

other areas, such as geothermal and ocean energy [8]. These developments aim to reduce the costs of deployment and enhance energy efficiency. Some studies are dedicated to inventing new methods or refining existing ones within energy facilities [20]. In particular systems, a combination of various renewable energies, like geothermal and solar, is employed to boost efficiency [16].

This document examines a geothermal energy system. Geothermal energy refers to the heat energy stored beneath the earth's surface [3]. Research indicates that the total heat emanating from the earth is approximately 42×10^{12} watts [3]. Of this energy, only 2% originates from the earth's crust. In comparison, 82% is produced in the mantle due to some physical aspects, and the rest is produced in the core. Despite the abundance of this energy source, its exploitation is confined to areas with suitable geological conditions [15].

At the start of the 21st century, the global capacity of geothermal energy reached nearly 9 GW [3], and at 2023, this value reached almost 15 GW [5]. This increase give an idea of the importance of this technology, used for electric energy generation and for non-electrical purposes includes a wide range of applications such as heat pumps, bathing, space heating, greenhouses, aquaculture, and industrial processes

A heat pump is used to transfer thermal energy from an external source to a building [14]. This external source may be either hot or cold, with the former typically yielding higher efficiency. For geothermal heat pumps, where the heat source is warm, the setup for the heat exchanger can be arranged in either horizontal or vertical layouts [11,19]. A vertical setup is generally more efficient since, at considerable depths, the ground's temperature remains fairly stable throughout the year, being cooler than the air in summer and warmer in winter. Conversely, horizontal configurations are installed closer to the surface, leading to more significant temperature fluctuations with the seasons. Despite the efficiency benefits of vertical installations, the higher costs associated with drilling boreholes often make the less expensive horizontal setup the preferred choice [6]. To enhance efficiency in horizontal systems, the exchanger is usually installed deeper underground [18].

Achieving optimal heat exchange performance necessitates precise knowledge of the geothermal system's functioning. Several works deals with the modeling and forecasting of geothermal systems. In [17] and [10], different Recurrent Neural Networs (RNN), Convolutional Neural Networks (CNN) and Deep Neural Networks (DNN) are applied to forecast the energy produced by a geothermal reservoir for big plants. In [1], different time series techniques and Time Delay Neural Networks are applied to a self-consumption geothermal system.

This study focuses on a geothermal heat exchanger installed in a bioclimatic home. The functioning of the installation has been monitored for over a year through temperature sensors placed throughout the exchanger.

The present research models the heat exchanger performance based on the temperature forecasting measured by a sensor located at the input collector connected to the heat pump. This would contribute to improve to energy planning, maintenance management and proper decision-making in general terms.

Three distinct approaches were examined following the dataset's preparation to determine which performed best. The outcomes were generally favourable, and accurate predictions of the heat exchanger's temperature input sensor's performance were made 24 h ahead of time. Additionally, it was determined that there is a significant correlation among various sensors distributed throughout the installation.

The structure of this document is organized in the following manner: After this introductory section, a concise description of the case study is provided. Subsequently, the section on methodologies employed is introduced, and the evaluation metrics. The subsequent section details the experiments, and then another one details the results and discussion. To conclude, the final section exposes the conclusions drawn and outlines directions for future research.

2 Case Study

This section describes the energy installation in which the study is carried out and the dataset description, visualization and preparation.

2.1 The Bioclimatic House

The ecological transition context described in the previous section enhances green energies that replace conventional fossil fuel technologies. One of the lines of action consists of promoting bioclimatic buildings that use different renewable sources to supply energy needs, reducing the environmental impact. However, this kind of building does not have a high presence in society, so different foundations and government administrations make significant investments to promote them. That is the case of the Sotavento Foundation, located in the north of the Galician region, Spain [21].

The geothermal installation consists of a five-hundred-meter pipe connected to a heat pump of 8.2 kW for obtaining Domestic Hot Water.

The geothermal facility is structured in three different parts, as shown in Fig. 1. The primary circuit consists of a horizontal heat exchanger whose main goal is to capture the ground heat and transfer it to the heat pump. This circuit has five different loops instrumented to measure temperature at different points. The heat pump represents the core unit of the system, receiving thermal energy from the primary circuit and sending it to the consumption point (secondary circuit).

2.2 Dataset Description

The dataset that serves as a case study in this document, namely the geothermal installation, comprises 30 features, 29 representing the different sensor values and another indicating the time label of the respective value capture, the temperature of S28 sensor. In Fig. 2, it can be seen how sensors 'S01' to 'S24' measure the temperature of different pipes with the idea of detecting possible blockages.

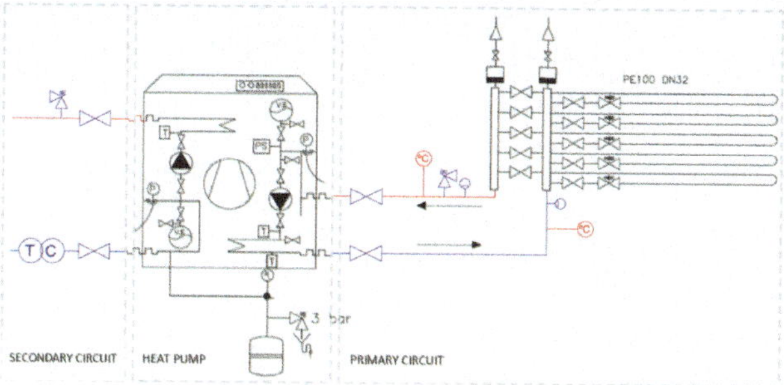

Fig. 1. Schematic of geothermal facility

Furthermore, 'S28' and 'S29' measures the temperatures at the input and output collector respectively, and 'S401' register the ground temperature. The dataset comprises 52.633 observations, with data collected every 10 min over one year.

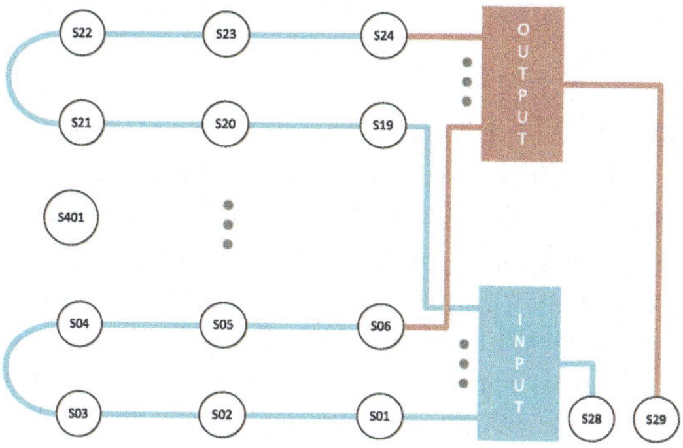

Fig. 2. Sensors located in the geothermal pipes and collectors

Data Analysis. For a visualization, the S28 values are plot grouping the one-year length values into days (Fig. 3a) and into hours (Fig. 3b) for presenting their respective observed variation range.

Table 1 details the description of the data. In short, the data is skewed, but with little variation given the mean values. Given that the mean is $\approx 17 \pm 5$, this means that most of the values are between 11.8 and 22. The distribution also has slightly flatter tails than a normal distribution, suggesting that fewer extreme values are reached.

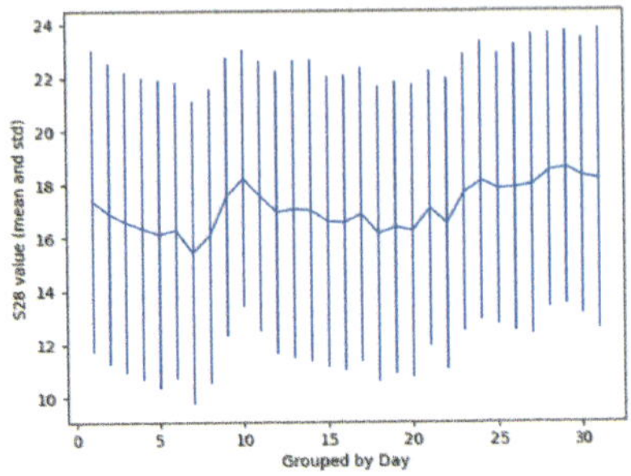

(a) Daily variation

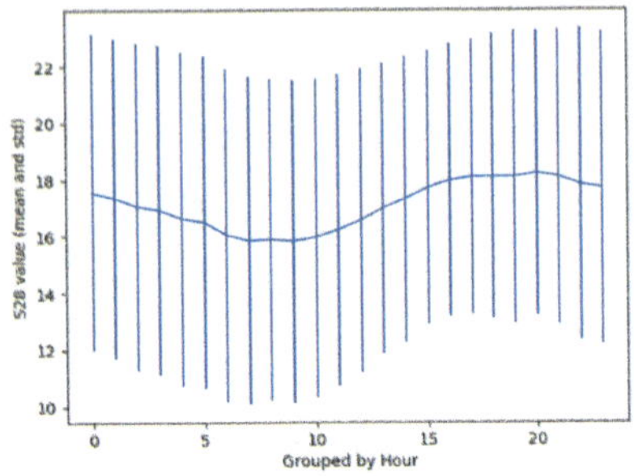

(b) Hourly variation

Fig. 3. Variation of the S28 value within one year. The horizontal line depicts the mean values and the vertical bars the standard deviation.

Table 1. Descriptive Statistics

Mean	StDev	Skewness	Kurtosis	Min	Median	Max
17.11	5.49	−0.58	−1.15	5.50	19.50	26.50

Data Preparation. The first step in the data preparation stage was to check for missing values and timestamps in the dataset under study. As this study focuses on a time series problem, the temporal sequence must not be broken. A meticulous dataset analysis verified that it did not present missing values, missing timestamps. However, some duplicate rows were identified, which were removed, leaving the dataset, at this point, with 52560 observations. Next, as the objective of this study is to forecast the S28 sensor's value for the next 24 h, it was necessary to group the data with an hourly frequency. As mentioned above, the original dataset presented a frequency of collection of values by the different sensors every 10 min. These were grouped into one-hour periods, aggregating the different sensors' values by their weighted average.

Additionally, the Gaussian distribution of the features included in the dataset was examined in the next stage. Verifying a normal distribution of data will allow the choice of methodology to be used in analysing the correlation of variables with the target of this study. The Kolmogorov-Smirnov test was used to verify this, with a p-value < 0.05. The results obtained from these tests made it possible to conclude that none of the features followed a normal distribution. With the distribution analysis pointing to a non-Gaussian distribution, Spearman's correlation coefficient test was used to analyse the correlation of the features, with a p-value < 0.05. This analysis is essential to select which features will feed the candidate models designed in this study. Considering our target, the S28 sensor, it was possible to verify that the features S29, S05, S12, S16, S17, S20, S21 and S401 were those that showed a strong correlation with the S28 values. Hence, all other features which did not correlate with S28 were removed.

Finally, as we are facing a supervised problem, it was necessary to create a sequence of data that depends on the number of timesteps given as input to the model to create an input sequence, X, with the corresponding label y. Furthermore, it was necessary to normalise the data, in the case of Multi Layer Perceptron (MLP) and CNNs, between 0 and 1, while in the case of Long Short-Term Memory (LSTMs), because this model works internally with the hyperbolic tangent, it normalises the values between -1 and 1.

3 Methods

The deep learning (DL) models used in this study are the MLP, LSTM and CNNs. These architectures are well known for their use in approaching time series forecasting problems [2,12,22].

In short, MLPs are feedforward neural networks that use supervised learning to map nonlinear relationships between inputs and outputs, and deal well with noise and missing values. LSTMs, on the other hand, are a type of recurrent neural network (RNN) that control how much information enters the cell, how much remains at the output to the next layer of the network, and how much is forgotten by the network. In CNNs, the convolution layer processes and extracts features, the pooling layer reduces the dimensionality of the data, and the fully connected layer finalises the predictions.

Additionally, two error metrics the Mean Absolute Error (MAE)

$$MAE = \frac{\sum_{i=1}^{n} |yi - yi_{pred}|}{n}$$

and the Root Mean Squared Error (RMSE)

$$RMSE = \sqrt{\frac{\sum_{i=1}^{n} (yi - yi_{pred})^2}{n}}$$

are employed to evaluate the effectiveness of the several candidate models (refer to [13]).

4 Experiments

Considering the objective of predicting the temperature of the S28 sensor for the next 24 h, several experiments were carried out with different LSTM, MLP and CNN-based candidate models. To obtain the best combination of hyperparameters, it was necessary to carry out a tuning process in several candidate models. To carry out these experiments, the data was divided into 70% for training and 30% for testing. Furthermore, a specific cross validator was applied, namely *TimeSeriesSplit*, with a k of 2.

All candidate models are evaluated in terms of performance by the two metrics mentioned in the previous section. Different hyperparameters with different values were considered in the developed tuning process. Table 2 summarizes the various hyperparameters and their respective range of values for the different DL models. All candidate models developed had the same seed value, in this case, 91195003. Regarding the technology we used, Python, version 3.9, was used to conceive, tune, and evaluate the different candidate models. Also, libraries like Numpy and scikit-learn, among others, were considered. For the development of DL models, TensorFlow v2.0.0 was used. Finally, all hardware used in the development of this study was provided by Google's Collaboratory.

Table 2. Hyperparameters' searching space.

Parameter	CNN	LSTM and MLP
Layers	[3, 4, 5]	[3, 4, 5]
Neurons/Filters	[32, 64, 128]	[16,32]
Activation	[ReLU, Tanh]	[ReLU, Tanh]
Timesteps	[12,24,36]	[12,24,36]
Batch Size	[5, 10, 20]	[5, 10, 20]
Dropout	[0.0, 0.5]	[0.0, 0.5]
Kernel Size	[3, 4, 5,6]	–
Pool Size	[2,3,4]	–

5 Results and Discussion

Table 3 summarises the results for the five best candidates for the three DL-conceived models, with their respective hyperparameters and training time.

The LSTM is the best performing model on this dataset, with an RMSE of 2.24 and an MAE of 1.72. Training required 24 time steps (one day) on an architecture composed of 64 neurons with *tanh* activation, a dropout rate of 0.5, and a batch size of 5. Data shown that architectures of three layers of 64 neurons with *tanh* activation and a batch size of 5 are well suited to this dataset.

Compared to MLP and CNN, the best RMSE for both models is below the top 5 LSTM results (with MAE just after), suggesting a significant difference in performance. In short, MLPs are homogeneous in terms of time steps and activations, with diversity in batch size, number of layers, number of neurons and dropout rate. The CNNs, on the other hand, are homogeneous in terms of neurons and activation, with varying values for the other parameters.

Overall, the analysis of the top-performing candidate models from the three deep learning models developed in this study revealed that the LSTM-based model exhibited the best performance. Furthermore, it was also found that the majority of the candidate models used a time step value of 24, with the activation function being another significant factor. The LSTM-based models utilized the *tanh* activation function, whereas the CNN-based and MLP-based models

Table 3. Top-5 results, per model. Legend: (a) timesteps; (b) batch size; (c) number of layers; (d) number of neurons/filters; (e) pool size; (f) kernel size; (g) dropout; (h) activation; (i) RMSE; (j) MAE; (k) time (s).

a	b	c	d	e	f	g	h	i	j	k
					LSTM candidate models					
24	5	3	64	–	–	0.5	tanh	2.24	1.72	583.28
36	20	4	64	–	–	0.0	tanh	2.37	1.82	718.58
24	5	3	64	–	–	0.0	tanh	2.38	1.92	364.96
12	20	3	128	–	–	0.5	tanh	2.44	1.98	599.78
36	5	3	32	–	–	0.0	tanh	2.50	2.00	631.61
					MLP candidate models					
24	5	4	64	–	–	0.0	ReLU	2.53	1.95	197.16
24	20	3	32	–	–	0.5	ReLU	2.62	2.19	95.46
24	5	3	32	–	–	0.0	ReLU	2.68	2.22	188.42
24	5	3	64	–	–	0.0	ReLU	2.69	2.25	238.40
24	5	3	32	–	–	0.5	ReLU	2.70	2.26	209.95
					CNN candidate models					
24	20	3	16	3	3	0.0	ReLU	2.79	2.25	121.51
36	20	3	16	3	3	0.0	ReLU	3.07	2.55	134.22
24	5	4	16	3	3	0.0	ReLU	3.17	2.78	369.96
24	20	3	16	2	5	0.0	ReLU	3.35	2.75	116.64
24	5	3	16	3	5	0.0	ReLU	3.39	2.83	298.59

employed the *ReLu* function. Notably, the best-performing MLP-based model required an additional layer, with a total of four layers, compared to the three layers used in the other two models.

6 Conclusions

In light of the unstable global economy and the issues posed by climate change, governments are placing a greater emphasis on renewable energy sources, including geothermal, wind, and solar power. Despite geological limitations, the adaptability of geothermal energy for both electrical and non-electrical purposes highlights its significance.

Hence, this study aimed to design, tune, and evaluate three DL models to forecast the S28 sensor' value over the next 24 h. For this purpose, several candidate models were designed based on LSTMs, CNNs and MLPs. Through various experiments, it was possible to verify that the best candidate model was LSTM-based, with an RMSE of 2.24 and a MAE of 1.72.

Despite all the experiences developed, there are more steps to take in future research. One of these steps is to use more DL models that deal with time series forecasting problems, such as Transformers, Gated Recurrent Units, and LST-Net, among others. Furthermore, hybrid models with the combination of models such as LSTMs and CNNs or Transformers and LSTMs should be used to understand whether using these types of models would result in better performance.

Acknowledgments. This work is financed by National Funds through the Portuguese funding agency, FCT - Fundação para a Ciência e a Tecnologia within project 2022.06822.PTDC (https://doi.org/10.54499/2022.06822.PTDC). The work of Pedro Oliveira was supported by the doctoral Grant PRT/BD/154311/2022 financed by the Portuguese Foundation for Science and Technology (FCT), and with funds from European Union, under MIT Portugal Program. CITIC is funded by the Xunta de Galicia through the collaboration agreement between the Consellería de Cultura, Educación, Formación Profesional e Universidades and the Galician universities for the reinforcement of the research centres of the Galician University System (CIGUS). This work has been supported by Centro Mixto de Investigación UDC-NAVANTIA (IN853C 2022/01), funded by GAIN (Xunta de Galicia) and ERDF Galicia 2021-2027.

References

1. Baruque, B., Porras, S., Jove, E., Calvo-Rolle, J.L.: Geothermal heat exchanger energy prediction based on time series and monitoring sensors optimization. Energy **171**, 49–60 (2019)
2. Cheker, Z., Chakkor, S., El Oualkadi, A.: Visual evoked potential classification support with convolutional neural network and recurrent neural network-a comparative study. In: 2022 International Conference on Decision Aid Sciences and Applications (DASA), pp. 1612–1617. IEEE (2022). https://doi.org/10.1109/DASA54658.2022.9765034
3. Dickson, M.H., Fanelli, M.: Geothermal Energy: Utilization and Technology. Routledge (2013)
4. (EC), E.C.: Europe 2020: A Strategy for Smart, Sustainable and Inclusive Growth. Working Paper (COM 2010) (2010)
5. Energy, G.: Data page: geothermal energy capacity. (2023). https://ourworldindata.org/grapher/installed-geothermal-capacity. Accessed 20 May 2024
6. Florides, G., Kalogirou, S.: Ground heat exchangers'a review of systems, models and applications. Renew. Energy **32**(15), 2461–2478 (2007). https://doi.org/10.1016/j.renene.2006.12.014
7. Hassan, Q., et al.: A comprehensive review of international renewable energy growth. Energy Built Environ. (2024)
8. Idroes, G.M., Hardi, I., Hilal, I.S., Utami, R.T., Noviandy, T.R., Idroes, R.: Economic growth and environmental impact: assessing the role of geothermal energy in developing and developed countries. Innov. Green Developm. **3**(3), 100144 (2024)
9. Igeland, P., Schroeder, L., Yahya, M., Okhrin, Y., Uddin, G.S.: The energy transition: the behavior of renewable energy stock during the times of energy security uncertainty. Renew. Energy **221**, 119746 (2024)
10. Jiang, A., Qin, Z., Faulder, D., Cladouhos, T.T., Jafarpour, B.: Recurrent neural networks for short-term and long-term prediction of geothermal reservoirs. Geothermics **104**, 102439 (2022). https://doi.org/10.1016/j.geothermics.2022.102439

11. Kakaç, S., Liu, H., Pramuanjaroenkij, A.: Heat Exchangers: Selection, Rating, and Thermal Design, Designing for Heat Transfer, 2nd edn. Taylor & Francis(2002)
12. Kiranyaz, S., Avci, O., Abdeljaber, O., Ince, T., Gabbouj, M., Inman, D.J.: 1d convolutional neural networks and applications: a survey. Mech. Syst. Signal Process. **151**, 107398 (2021). https://doi.org/10.1016/j.ymssp.2020.107398
13. Naser, M., Alavi, A.H.: Error metrics and performance fitness indicators for artificial intelligence and machine learning in engineering and sciences. Architect. Struct. Construct. **3**(4), 499–517 (2023). https://doi.org/10.1007/s44150-021-00015-8
14. Ozgener, L.: A review on the experimental and analytical analysis of earth to air heat exchanger (EAHE) systems in turkey. Renew. Sustain. Energy Rev. **15**(9), 4483–4490 (2011). https://doi.org/10.1016/j.rser.2011.07.103
15. Ozgener, L., Ozgener, O.: Monitoring of energy exergy efficiencies and exergoeconomic parameters of geothermal district heating systems (GDHSS). Appl. Energy **86**(9), 1704–1711 (2009). https://doi.org/10.1016/j.apenergy.2008.11.017
16. Ozgener, O., Ozgener, L.: Modeling of driveway as a solar collector for improving efficiency of solar assisted geothermal heat pump system: a case study. Renew. Sustain. Energy Rev. **46**, 210–217 (2015). https://doi.org/10.1016/j.rser.2015.02.043
17. Puppala, H., Saikia, P., Kocherlakota, P., Suriapparao, D.V.: Evaluating the applicability of neural network to determine the extractable temperature from a shallow reservoir of puga geothermal field. Int. J. Thermofluids **17**, 100259 (2023). https://doi.org/10.1016/j.ijft.2022.100259
18. Rezaei, A., Kolahdouz, E., Dargush, G., Weber, A.: Ground source heat pump pipe performance with tire derived aggregate. Int. J. Heat Mass Transf. **55**(11–12), 2844–2853 (2012). https://doi.org/10.1016/j.ijheatmasstransfer.2012.02.004
19. Sauer, H., Howell, R.: Heat Pump Systems. Krieger Publishing Company (1991)
20. Sayed, E.T., Olabi, A.G., Alami, A.H., Radwan, A., Mdallal, A., Rezk, A., Abdelkareem, M.A.: Renewable energy and energy storage systems. Energies **16**(3), 1415 (2023)
21. Sotavento. (2023). https://www.sotaventogalicia.com/area-tecnica/instalaciones-renovables/minieolica/. Accessed 13 Jan 2024
22. Zhao, Z., Chen, W., Wu, X., Chen, P.C., Liu, J.: LSTM network: a deep learning approach for short-term traffic forecast. IET Intel. Transport Syst. **11**(2), 68–75 (2017). https://doi.org/10.1049/iet-its.2016.0208

A Hybrid Intelligence Model Forecasts the Temperature of a Battery Used in Electric Vehicles

Manuel Rubiños[1], Paula Arcano-Bea[1], Antonio Díaz-Longueira[1(✉)], Álvaro Michelena[1], Rafael Vega Vega[1], José-Luis Casteleiro-Roca[1], and José Manuel Andújar[2]

[1] CTC, Department of Industrial Engineering, CITIC, University of A Coruña, EPEF, Calle Mendizábal s/n, Campus de Esteiro, 15403 Ferrol, A Coruña, Spain
{manuel.rubinos,paula.arcano,a.diazl,alvaro.michelena, rafael.alejandro.vega.vega,jose.luis.casteleiro}@udc.es

[2] Department of Electronic Engineering, Computer Systems and Automatic, University of Huelva, Campus de El Carmen, 21071 Huelva, Huelva, Spain
andujar@diesia.uhu.es

Abstract. Battery technology is advancing due to the requirement to store reliable energy in electric transportation systems and portable electronics devices. Assessing battery conditions and increasing the autonomy of electric vehicles requires to improve the behavior models. Prognosticating battery performance, nevertheless, is complex due to multiple variables, such as temperature, wear, and usage guidelines. Understanding battery performance requires a depth knowledge of, for example, time series analysis, above all forecasting. This research uses Artificial Neural Networks (ANN) modeled taking into account system dynamics, to create a Lithium Ion Battery (LIB) prediction model. Testing was performed with a 59.2 VDC, 120 Ah LIB, monitoring voltage, current and temperature while simulating charges and discharges. The model used is a hybrid type, which uses clustering to generate internal local models. This model predict the temperature development with very good results in terms of Mean Squared Error. From this research it is deduced that hybrid models are appropriate tools for prognostic battery performance.

Keywords: Artificial intelligent · Batteries · Electric vehicles · Temperature prediction

1 Introduction

The increase in popularity of electronic devices demands to store energy safe and efficient during the last times. Electric Vehicles (EVs) and portable electronics like laptops and phones, [16] manifests that battery technology and associated devices will continue their development.

For this moment, the transport sector is replacing the use of coal for the Lithium Ion Batteries (LIBs) [15]. Also safe models, control methods to expand

F. Zayas-Gato et al. (Eds.): DCAI 2024, LNNS 1173, pp. 52–62, 2024.
https://doi.org/10.1007/978-3-031-73910-1_6

EV autonomy should be used to estimate the autonomy of these vehicles [19]. On the sidelines of the operating conditions of the battery, it should always use a safe model to establish an effective control system. For a given set of conditions and power demands, the model must be trained to reliably predict battery conditions and, at the same time, its remaining operating time and its energy availability. But factors such as time, charge and discharge cycles, temperature, wear, and also the demand for energy required instantly by the batteries affects their performance [17]. Providing an effective, reliable and computationally economical solution to the proposed problem for each immediate battery operation situation requires modeling solutions that can adjust or modify the dynamic of the model.

The analysis and modeling of failures of these batteries is mainly motivated by their high application in portable devices such as EVs and by safety requirements [4], clustering [1], classification [12], query by content [2], anomaly detection [5], and forecasting [11]. On the other hand, deep learning is now an essential instrument of the new generation of time series forecasting models, which minimize the problems of traditional machine learning and are more powerful at missing values, are capable of identifying highly complex patterns in data and are effective in long-term forecasts [7]. This is happening due to the recent increase in data availability and processing power.

Time series forecasting problems with variable length of input sequences, connected handwriting recognition [9] or speech recognition [14], are the most common architectures [8]. The developments known as Long Short-Term Memory (LSTM) [10] and Gated Recurrent Units (GRU) [6] were conceived to overcome the impediment of gradient disappearance that arises during training of conventional RNNs.

The integration of hybrid models that combine intelligent algorithms and deep learning techniques has proven effective in accurately predicting the State Of Charge (SOC) and State Of Health of electric vehicle batteries. Recent research has applied networks neural networks informed by physics and digital twins based on machine learning to improve the estimation and management of the life cycle of these batteries, optimizing their performance and safety under various operating conditions [13,20].

The goal of this research is to implement a Hybrid Intelligent Model to predict the temperature of a power cell, using a combination of clustering an regression techniques based on Artificial Neural Networks (ANNs). The selection of the strategy is made based on the minimum Mean Square Error (MSE) criterion.

Section 2 presents the case study used in the experiments. Section 3 presents the basic model approach, while Sect. 4 discusses the obtained results. At the end, some conclusions and future work are provided.

2 Case Study

A 59.2 VDC, 120 Ah LIB was subjected to two experiments (charging and discharging tests) to estimate and validate the adaptive capacity of the algorithm created to model the LIB. The battery studied has the following properties: Aluminum box + BMS 80 A + Samsung 50E lithium-ion pack. Samsung 50E Lithium-ion Battery Specifications are 59.2V, 120Ah., NMC technology. The nominal voltage of the battery pack is 59.2 V. The voltage for charging the battery packs is 67.2 V; 120 Ah (7.10 kWh) of capacity. Maximum continuous discharge current: 240 A. To reach 120 Ah in this case, the assembly requires 16 cells of 3.7 V in series and 24 cells of 5 Ah in parallel. These cells, specially designed and manufactured by Batesur©, are Samsung model INR21700-50S cells. To do so, experimental data were gathered on the actual battery using a dedicated National Instruments DAQ with a sampling time fixed to one second. The sample time was selected taking into account previous attempts made with this experimental equipment and was configured in the SCADA developed in LabVIEW. It is necessary to comment that in the previous tests, the authors try to reduce this sampling time to acquire signals more quickly, but the change in the recorded values is not significant.

The experiments were performed in a control environment, that is, its operating temperature was controlled by a thermal chamber. More precisely, during the two tests several heating cycles were programmed up to a maximum internal temperature of 60 °C, followed by forced cooling of the battery taking into account the manufacturer's safety temperature restrictions. High flow axial fans were used during the cooling cycles. To understand how temperature affects voltage in the model training and validation stages, the goal of temperature control was not to reach a certain temperature, but to cause significant temperature fluctuations over a short period of time.

For the experiments, a 32 kW programmable source and sink from the Regatron© TC.GSS series were used. The defined current or power limits and the setpoint voltage setting determine whether the device can operate in the first (source) or fourth (load) quadrant. To constitute the electrical connection between the programmable source and the battery, direct connection based on fuses was used. It was not necessary to add antireverse diodes or other additional components because the device already has built-in protections. The charge and discharge profiles (current and power profiles, respectively), which were then carried out directly in the final assembly, were defined with the programmable source software. To measure the variables of interest such as voltage, current and battery operating temperature, an ad hoc signal processing circuit and specific sensors were used. The Hall effect sensors used were the LEM© model LV25-P and LA25-P to measure voltage and current, respectively. To measure the temperature it was used an NTC type thermoresistance and a conditioner circuit made using a resistive divider and a zero correction circuit. The temperature sensor was placed at the other end of the heat source of the thermal chamber, in the center of the side face of the battery, to ensure an accurate and uniform result. The case study is described deeply in [3], and also explain the dataset used in this research.

In addition, a contact thermometer was used as a reference to compare the recorded measurements. To capture the variables, was used a National Instruments© NI-USB-6008 data acquisition (DAQ) board with 12-bit resolution. For their storage, processing, and representation, was used the software ad hoc developed LabVIEW© programmed data collection and control system. The experimental setup used is shown in Fig. 1.

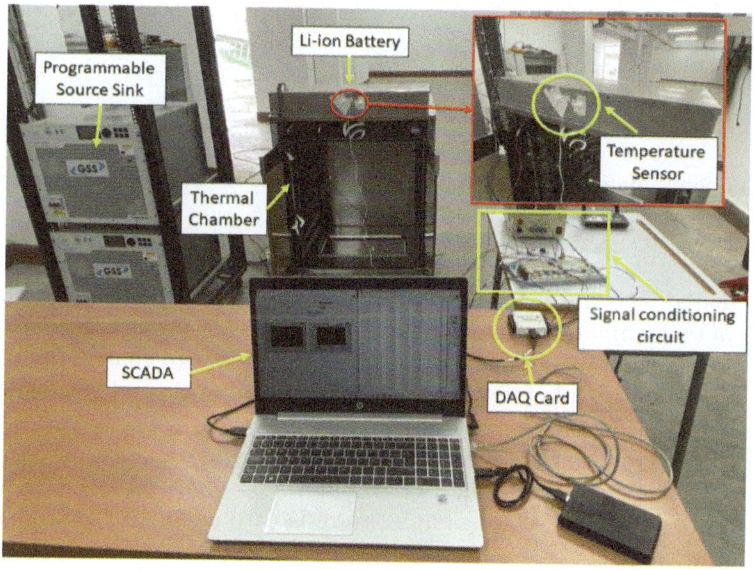

Fig. 1. Experimental setup in a laboratory.

2.1 Model Approach

In this research, a model was created with the objective of predicting the evolution of the Temperature variable measured by the NTC mentioned in the previous section. Figure 2 represents a simplified data flow diagram of the model, showing the 3 inputs that the model uses to predict the Temperature difference. These three inputs are the Current and Voltage at an instant in which the output is calculated, and also the Temperature at the previous instant.

The model used in this research is a hybrid model, and it is internally divided into several local models. Only one of these local models is used at a time; depending on the model inputs, the specific local model is chosen to calculate the output. The use of this type of models is justified to optimize the performance of local models; these individual models have the ability to better tailor the output for a specific part of the data set.

Additionally, the model predicts the Temperature difference instead of predicting the Temperature directly, since the Temperature is an input to the model. In models with very similar values between any input and the desired variable

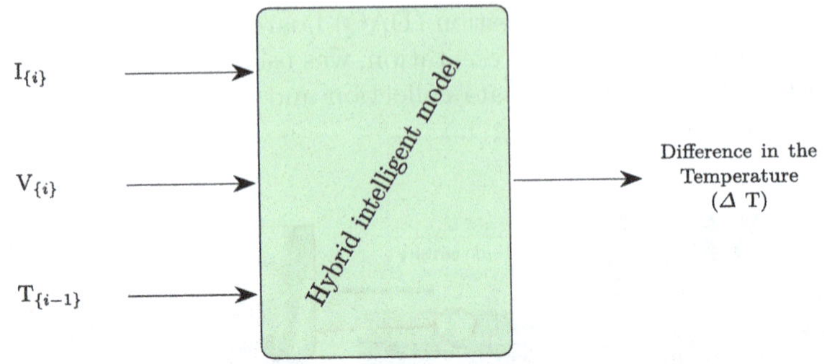

Fig. 2. Simplified data-flow diagram to predict difference in the Temperature in the battery.

to be predicted, the training phase is used to make the output the same as the similar input: the models obtained in these cases do not predict the desired variable. When this fact appears, a simple way to avoid it is to change the predicted variable by a difference between one instant and the next, and this is what was done in this research.

Some of the samples were isolated to perform a validation test once the model was created. The rest of the dataset was normalized to perform the training phase, however, another set of samples were isolate again to use to decide the hybrid topology.

3 Results and Discussion

The results of this research are divided in four different subsections; first of all, there is presented the clustering result for the model created and then, the local models results are shown. The next two subsections described the hybrid model results, and the validation of the model.

3.1 Clustering Results

Table 1 shows the clusters created to predict the Temperature variation from one instant to the next one. It is necessary to emphasize that the hybrid model developed is based in the division of the whole dataset in several clusters. As the optimal division is not known, there were made up to ten different topologies of the system, the global model and nine hybrid systems with different number of clusters. Table 1 shows the number of samples for each cluster when the dataset is divided using K-Means. As the K-Means algorithm is used to create the clusters, the unique parameter that need is the number of desire clusters. There where created 9 different hybrid topologies, the dataset was divided several times with the aim to create 2, 3, 4, etc. until 9 clusters.

The results shown in Table 1 are organized in columns; each column represent a complete model, and each row represent the local model results for each cluster

Table 1. Number of samples for each cluster

	Global	Hyb. 2	Hyb. 3	Hyb. 4	Hyb. 5	Hyb. 6	Hyb. 7	Hyb. 8	Hyb. 9	Hyb. 10
C-1	93520	45554	10981	10831	8351	7040	3650	3650	1869	1902
C-2		47966	38766	21017	10780	10556	6765	6393	3651	3645
C-3			43773	22554	16596	13586	6991	6765	6217	4184
C-4				39118	20297	18504	13590	12040	6765	5571
C-5					37496	21275	18813	13144	11828	6765
C-6						22559	21155	13854	11956	11055
C-7							22556	16699	13624	11659
C-8								20975	16641	13429
C-9									20968	14565
C-10										20746

in the model. As the first column represent the global model, it only has one cluster (one row); the last column present the results for a hybrid model with 10 clusters (it needs to used the ten rows).

3.2 Local Regression Models

A total of 17 regression algorithms are tested in each created cluster. 15 ANNs, with a feedforward configuration, one hidden layer with tan-sigmoid activation function, and a linear function in the output layer. The number of neurons in the hidden layer was varied from 1 to 15. Moreover, polynomial regression algorithm is trained for the clusters data; first and second degree is chosen in this research as different algorithms to test.

Table 2 shows the MSE (Mean Squared Error) as a performance measurement of the regression algorithm. Considering that the data was normalized prior to training phase, and the values of the MSE in the tables are in 10^{-3} range (it is indicated in the caption of the table). The MSE values are obtained used K-fold cross validation; this is a validation techniques that trains several times the same algorithm to use all available data to calculated the performance. K-fold cross validation divides the data 10 times (in this specific researcher case) and uses each group as test dataset (and the rest of the groups as train dataset). This validation techniques requires that each algorithm was trained k times, 10 times in this paper.

Table 3 indicates the best regression technique and its configuration for each local model. The "ANN*" means an Artificial Neural Network with * neurons in the hidden layer.

3.3 Hybrid Model Results

To check what is the best internal topology of the final model, a dataset isolated from the local models training, is used to calculate the performance of

Table 2. MSE values for each best local model ($\cdot 10^{-3}$)

	Global	Hyb. 2	Hyb. 3	Hyb. 4	Hyb. 5	Hyb. 6	Hyb. 7	Hyb. 8	Hyb. 9	Hyb. 10
C-1	0.4554	0.3235	0.0088	0.0091	0.9264	1.0567	0.0136	0.0134	0.0043	0.0043
C-2		0.4492	0.2513	0.2558	0.0091	0.0089	0.0055	1.0990	0.0130	0.0134
C-3			0.5888	0.3025	0.0202	0.0082	1.0047	0.0055	1.1461	0.6730
C-4				0.5687	0.2875	0.1993	0.0083	0.4718	0.0055	1.0375
C-5					0.5450	0.7932	0.2160	0.0084	0.0088	0.0056
C-6						0.2801	0.7861	0.2151	0.5004	0.0091
C-7							0.2876	0.2045	0.2152	0.4844
C-8								0.6944	0.1967	0.2123
C-9									0.6963	0.1298
C-10										0.6695

Table 3. Best local model algorithm

	Global	Hyb. 2	Hyb. 3	Hyb. 4	Hyb. 5	Hyb. 6	Hyb. 7	Hyb. 8	Hyb. 9	Hyb. 10
C-1	ANN15	ANN15	ANN15	ANN15	ANN13	ANN15	ANN15	ANN12	ANN4	ANN1
C-2		ANN15	ANN13	ANN13	ANN15	ANN14	ANN15	ANN13	ANN13	ANN15
C-3			ANN14	ANN15	ANN4	ANN14	ANN14	ANN12	ANN12	ANN9
C-4				ANN14	ANN14	ANN15	ANN15	ANN15	ANN12	ANN14
C-5					ANN15	ANN15	ANN9	ANN15	ANN11	ANN13
C-6						ANN15	ANN13	ANN10	ANN13	ANN14
C-7							ANN15	ANN5	ANN15	ANN12
C-8								ANN15	ANN5	ANN11
C-9									ANN12	ANN12
C-10										ANN14

each topology. In this phase, all the internal models are trained with the configuration selected in the previous phase. As it is mentioned before, the data was normalized, and the values shows for the MSE are in range of 10^{-3} (it is referenced in the caption of the table). The results of this test is shown in Table 4.

Table 4. MSE values for each hybrid configuration ($\cdot 10^{-3}$)

Global	Hyb. 2	Hyb. 3	Hyb. 4	Hyb. 5	Hyb. 6	Hyb. 7	Hyb. 8	Hyb. 9	Hyb. 10
0.4245	0.3533	0.3613	0.3510	0.3519	0.3410	0.3540	**0.3276**	0.3419	0.3332

In Table 5 the best algorithms for the final local model are shown. The training of the final model requires that the final local models must the trained one more time. As cross validation is used to calculate the performance of the local models, the previous regression models didn't use all the available data for each cluster. Once the final topology is defined, new models were created using all the data in the final clusters topology.

Table 5. Algorithm for each local model inside the hybrid topology

Cluster 1	Cluster 2	Cluster 3	Cluster 4	Cluster 5	Cluster 6	Cluster 7	Cluster 8
ANN12	ANN13	ANN12	ANN15	ANN15	ANN10	ANN5	ANN15

3.4 Validation Results

Once the hybrid models are chosen, a validation test was performed to calculate the accuracy of each model using a different dataset. Table 6 shows the performance metrics for the hybrid model created. In this table the MSE, the MAE (Mean Absolute Error) and the NMSE (Normalized MSE) calculated with raw data and with normalized data [18].

Table 6. Performance metrics for the best hybrid configuration

	MSE	MAE	NMSE
Hybrid configuration with 10 local models	0.1739	0.1380	1.8317
(with normalized values)	$4.3339 \cdot 10^{-4}$	0.0060	

Figure 3 show the error plot for the validation dataset for the model. The upper plots are in real values, and the lower plots are created with normalized values.

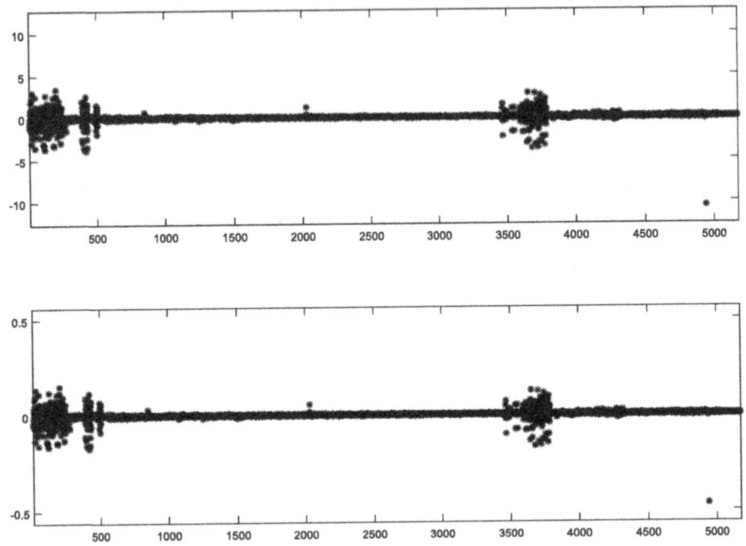

Fig. 3. Error plot (upper with real values, °C, lower with normalized values)

4 Conclusions and Future Work

The main conclusion of this research is that it is possible to use hybrid topologies to model the behavior of batteries. A model was created to predict the Temperature difference (ΔT) of a battery. The performance of the prediction of the Temperature obtained values lower than $138 \cdot 10^{-3}$ °C.

The model created has a hybrid topology, with 8 internal local model; all of them use Artificial Neural Network as regression technique. The training procedure uses K-fold cross validation with 10 folds, that ensure more realistic error values than used hold out to calculated the training error.

The authors will try to continue this research with the creation of new models to predict other variables of the system like the State Of Charge of the battery.

Acknowledgements. This work has been supported by Xunta de Galicia through Axencia Galega de Innovación (GAIN) by grant IN853C 2022/01, Centro Mixto de Investigación UDC-NAVANTIA "O estaleiro do futuro", which is ongoing until the end of September 2025. The support was inherited from both the starting and consolidation stages of the same project throughout 2015-2018 and 2018-2021, respectively. This stage is also co-funded by ERDF funds from the EU in the framework of program FEDER Galicia 2021-2027.

Álvaro Michelena's research was supported by the Spanish Ministry of Universities (https://www.universidades.gob.es/), under the "Formación de Profesorado Universitario" grant with reference FPU21/00932.

Antonio Díaz-Longueira's research was supported by the Xunta de Galicia (Regional Government of Galicia) through grants to Ph.D. (http://gain.xunta.gal), under the "Axudas á etapa predoutoral" grant with reference: $ED481A2023072$.

This research is co-financed by the Interreg Atlantic Area Programme through the European Regional Development Fund, *EAPA_0019/2022* SAtComm project.

Xunta de Galicia. Grants for the consolidation and structuring of competitive research units, GPC (ED431B 2023/49).

CITIC, as a center accredited for excellence within the Galician University System and a member of the CIGUS Network, receives subsidies from the Department of Education, Science, Universities, and Vocational Training of the Xunta de Galicia. Additionally, it is co-financed by the EU through the FEDER Galicia 2021-27 operational program (Ref. ED431G 2023/01).

References

1. Aghabozorgi, S., Shirkhorshidi, A.S., Wah, T.Y.: Time-series clustering-a decade review. Inf. Syst. **53**, 16–38 (2015)
2. Agrawal, R., Faloutsos, C., Swami, A.: Efficient similarity search in sequence databases. In: Lomet, D.B. (ed.) FODO 1993. LNCS, vol. 730, pp. 69–84. Springer, Heidelberg (1993). https://doi.org/10.1007/3-540-57301-1_5
3. Andújar, J.M., Barragán, A.J., Vivas, F.J., Enrique, J.M., Segura, F.: Iterative nonlinear fuzzy modeling of lithium-ion batteries. Batteries **9**(2), 100 (2023)
4. Antón, J.C.Á., Nieto, P.J.G., de Cos Juez, F.J., Lasheras, F.S.S., Viejo, C.B., Gutiérrez, N.R.: Battery state-of-charge estimator using the mars technique. IEEE Trans. Power Electron. **28**(8), 3798–3805 (2012)
5. Chen, C.W., Chiu, L.: Ordinal time series forecasting of the air quality index. Entropy **23**(9), 1167 (2021)
6. Cho, K., et al.: Learning phrase representations using RNN encoder-decoder for statistical machine translation. arXiv preprint arXiv:1406.1078 (2014)
7. Ciregan, D., Meier, U., Schmidhuber, J.: Multi-column deep neural networks for image classification. In: 2012 IEEE Conference on Computer Vision and Pattern Recognition, pp. 3642–3649. IEEE (2012)
8. Dupond, S.: A thorough review on the current advance of neural network structures. Annu. Rev. Control. **14**(14), 200–230 (2019)
9. Graves, A., Liwicki, M., Fernandez, S., Bertolami, R., Bunke, H., Schmidhuber, J.: A novel connectionist system for improved unconstrained handwriting. IEEE Trans. Pattern Anal. Mach. Intell. **31**(5), 855–868 (2009)
10. Hochreiter, S., Schmidhuber, J.: Long short-term memory. Neural Comput. **9**(8), 1735–1780 (1997)
11. Hu, J., Niu, H., Carrasco, J., Lennox, B., Arvin, F.: Voronoi-based multi-robot autonomous exploration in unknown environments via deep reinforcement learning. IEEE Trans. Veh. Technol. **69**(12), 14413–14423 (2020)
12. Keogh, E., Kasetty, S.: On the need for time series data mining benchmarks: a survey and empirical demonstration. In: Proceedings of the eighth ACM SIGKDD International Conference on Knowledge Discovery and Data Mining, pp. 102–111 (2002)
13. Kumari, P., Kumar, N.: Hybrid optimized deep learning approach for prediction of battery state of charge, state of health and state of temperature. Electric. Eng. **106**(2), 1283–1290 (2024)
14. Li, X., Wu, X.: Constructing long short-term memory based deep recurrent neural networks for large vocabulary speech recognition. In: 2015 IEEE International Conference on Acoustics, Speech and Signal Processing (ICASSP), pp. 4520–4524. IEEE (2015)

15. Lowe, I.: How planning can address the challenge of transitioning to low-carbon urban economies. In: The Routledge Handbook of Australian Urban and Regional Planning, pp. 199–208. Routledge (2017)
16. Philippot, M., Alvarez, G., Ayerbe, E., Van Mierlo, J., Messagie, M.: Eco-efficiency of a lithium-ion battery for electric vehicles: influence of manufacturing country and commodity prices on ghg emissions and costs. Batteries **5**(1), 23 (2019)
17. Ren, D., Hsu, H., Li, R., Feng, X., Guo, D., Han, X., Lu, L., He, X., Gao, S., Hou, J., et al.: A comparative investigation of aging effects on thermal runaway behavior of lithium-ion batteries. ETransportation **2**, 100034 (2019)
18. Saigal, S., Mehrotra, D.: Performance comparison of time series data using predictive data mining techniques. Adv. Inf. Mining **4**(1), 57–66 (2012)
19. Tehrani, K.: A smart cyber physical multi-source energy system for an electric vehicle prototype. J. Syst. Architect. **111**, 101804 (2020)
20. Xu, L., Wu, F., Chen, R., Li, L.: Data-driven-aided strategies in battery lifecycle management: prediction, monitoring, and optimization. Energy Storage Mater. **59**, 102785 (2023)

A Comparative Analysis of Algorithms and Metrics to Perform Clustering

Manuel Rubiños[1], Antonio Díaz-Longueira[1], Míriam Timiraos[1,2], Álvaro Michelena[1(✉)], María Teresa García-Ordás[3], and Héctor Alaiz-Moretón[3]

[1] CTC, Department of Industrial Engineering, CITIC, University of A Coruña, EPEF, Calle Mendizábal s/n, Campus de Esteiro, 15403 Ferrol, A Coruña, Spain
{manuel.rubinos,a.diazl,miriam.timiraos.diaz,alvaro.michelena}@udc.es
[2] Fundación Instituto Tecnológico de Galicia, Department of Water Technologies, National Technological Center, 15003 A Coruña, Spain
[3] Department of Electrical and Systems Engineering, University of León, Escuela de Ingenierías Industrial, Informática y Aeroespacial, Campus de Vegazana, s/n, 24071 León, Spain
{m.garcia.ordas,hector.moreton}@unileon.com

Abstract. This study introduces a novel approach in the field of soft computing, focused on determining optimal clustering algorithms and evaluation criteria for datasets. Using six complex datasets and MatLab R2023a software, especially the `evalclusters` function, various clustering algorithms such as k-means, agglomerative clustering and Gaussian mixture distribution are evaluated, along with evaluation criteria such as Calinski-Harabasz index, Davies-Bouldin index, gap criterion and silhouette metrics.

The methodology focuses on selecting the best algorithm and evaluation criteria to perform automatic clustering of a dataset. It is experimented with two- and four-cluster datasets, challenging due to their specific distributions. The procedure includes data loading, elimination of pre-existing clustering assignments, and the use of MatLab's `evalclusters` function to determine the optimal number of clusters and sample assignment, considering a range of clusters to be evaluated.

The results highlight how different combinations of algorithms and evaluation criteria handle the clustering complex data, with specific examples showing both effective and ineffective clustering. Comparative tables are presented for each dataset, highlighting the combinations of algorithms and criteria that achieve accurate clustering.

In conclusion, combinations of algorithms and criteria are identified that work effectively on most datasets, finally recommending the use of k-means with the Calinski-Harabasz index to obtain the optimal number of clusters. This study not only validates previous findings, but also extends the applications of soft computing in the classification of complex data sets.

Keywords: Agglomerative · Calinski-Harabasz · Davies-Bouldin · Gap criterion · Gaussian mixture distribution · *k*-means · Silhouette

F. Zayas-Gato et al. (Eds.): DCAI 2024, LNNS 1173, pp. 63–72, 2024.
https://doi.org/10.1007/978-3-031-73910-1_7

1 Introduction

In the field of soft computing, the exploration of innovative techniques for analyzing and interpreting complex datasets has increased attention. It opens the possibility of analyzing different data classification way from another point of view. In previous works like [1,2,4,5], the authors created clusters from a dataset in an arbitrary way, and after that they decided the optimal number of groups depending on the accuracy of the created models.

The present study introduces a novel approach to determine the optimal clustering algorithms and evaluation criteria for complex datasets. By leveraging six distinct datasets characterized by their intricate distributions, this research underscores the importance of methodical algorithm selection and criterion evaluation in the realm of soft computing. Utilizing the advanced functionalities of MatLab R2023a, specifically the `evalclusters` function, the study meticulously evaluates various clustering algorithms and criteria, including k-means, agglomerative clustering, Gaussian mixture distribution, along with the Calinski-Harabasz index, Davies-Bouldin index, gap criterion, and silhouette metrics.

Through an exhaustive experimental procedure, the research systematically identifies the most efficacious combinations of algorithms and criteria, aiming to enhance clustering accuracy and reliability. The results and discussions presented herein not only corroborate the findings of previous studies, but also push the boundaries of soft computing applications, setting the stage for future innovations in this dynamic field, making the selection of clusters number automatically.

After this introduction, Sect. 2 presents the Methodology used in this research. Section 3 presents the Results and the discussion; and, lastly, in Sect. 4, the Conclusions and future works are offered.

2 Methodology

As the main aim of this research is to determine the best algorithm and metric to perform an automatic clustering of a dataset, six specific datasets [3] were utilized to analize the optimal decision for clustering. The original datasets are shown in Fig. 1; they are made up of two dimensional data with 2 and 4 clusters. These datasets were chosen because they contain complex data for clustering performance due to their specific distribution.

The software used to perform the research is MatLab R2023a, which includes a specific function that simplified the task of finding the optimal number of clusters.

As `evalclusters` allows to set several parameters, this research uses the function with the next syntax: `evaluation = evalclusters(x, clust, criterion, Name, Value)`.

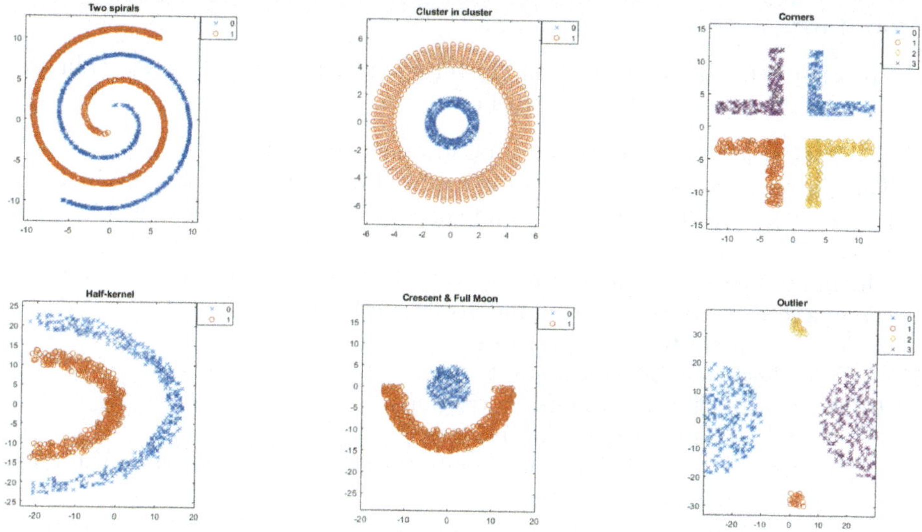

Fig. 1. Original datasets.

- Dataset to divide in clusters (`x`).
- Clustering algorithms (`clust`)
  - k-means clustering - `'kmeans'`
  - Agglomerative clustering - `'linkage'`
  - Gaussian mixture distribution algorithm - `'gmdistribution'`
- Clustering evaluation criterion (`criterion`)
  - Calinski-Harabasz index - `'CalinskiHarabasz'`
  - Davies-Bouldin index - `'DaviesBouldin'`
  - Gap criterion clustering evaluation object - `'gap'`
  - Silhouette values - `'silhouette'`
- Options (`Name,Value`):
  - List of number of clusters to evaluate - `'KList',[min:max]`

The output of the function has different information, such as like the optimal number of clusters, the assignation for each sample to its cluster and the criterion values. Even though there are more information in the output of the function, this research only uses the optimal number of clusters and the assignation of the dataset.

In order to allow the assignation of new data to the clusters, it is necessary to know the centroids of each clusters to measure the distance between the sample and the centroids, and assign the closest cluster. The `evalclusters` function does not have the centroids of the optimal clusters, so it is mandatory to create new clusters with the chosen algorithm once the optimal number of clusters is known.

Procedure

The procedure for choosing the best clustering algorithm and the best evaluation criterion is described in the next workflow:

1. The data is loaded using the scripts downloaded from [3].
2. The cluster assigned to each data is removed from the data: the original data is a matrix with three columns, and the third one is removed (the clustering label).
3. The minimum and maximum number of clusters is defined to be an input of the `evalclusters` function (3 and 10 respectively).
4. Repeat for all clustering algorithms.
   (a) The clustering algorithm is set.
   (b) Repeat for all the evaluation criterions.
      i. The evaluation criterion is set.
      ii. Repeat until the optimal number of clusters is lower than the maximum number of clusters.
         - The optimal number of clusters is calculated throw the function.
         - Increment the maximum number of clusters.
      iii. Plot a figure with the clustering assignation.

After several tests with different number of minimum and maximum number of clusters, the minimum one is set to 3, and the initial maximum to 10, as it is mentioned in the workflow. It is necessary to remark that the maximum number of clusters is not fixed, so if a result gives the same number for both maximum and optimal, this evaluation is not accepted. It is not really known if this is the best number of clusters, or the algorithm returns this value as it is not allowed to use more clusters. When this situation happens, the maximum is increased by one and the function `evalclusters` is executed again. Figure 2 shows a diagram that represents the workflow.

3 Results and Discussion

The results of this research focus on the analysis of the clusters of data created with the previously described procedure. The main aim of the analysis is to check if the data from the different original clusters data is mixed or not.

As an example, Fig. 3 shows the clustering results after running the k-means algorithm using Calinski-Harabasz index to calculate the optimal number of clusters for Cluster in Cluster data. It can be seen that, despite having the original data only two clusters, the algorithm finally clusters the data in a good way as none of the internal data is mixed with outside data, despite the fact that it uses more clusters.

Another example can be seen in Fig. 4. The figure shows the results after running Gaussian mixture distribution algorithm using Gap criterion clustering evaluation to calculate the optimal number of clusters for Half-Kernel data. The three final clusters have mixed data between the two original clusters, which means that the data is not clustered in a good way.

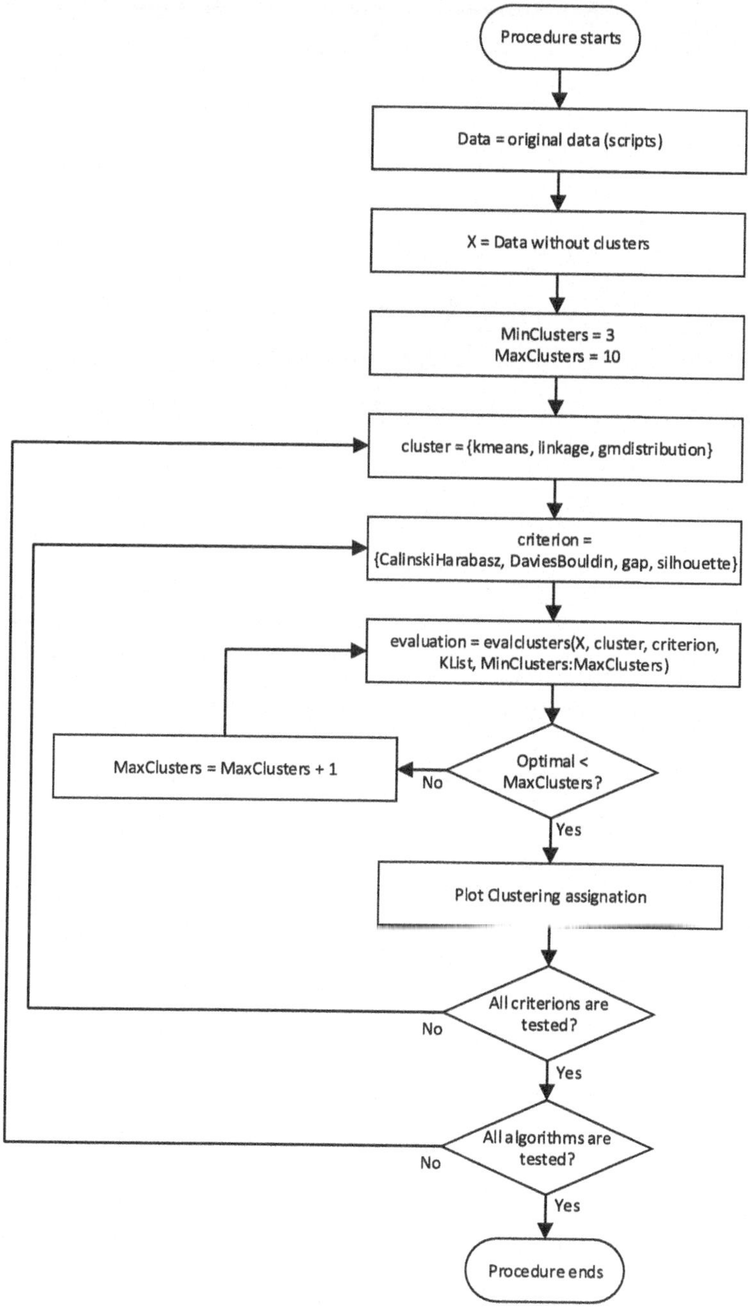

Fig. 2. Workflow of the procedure.

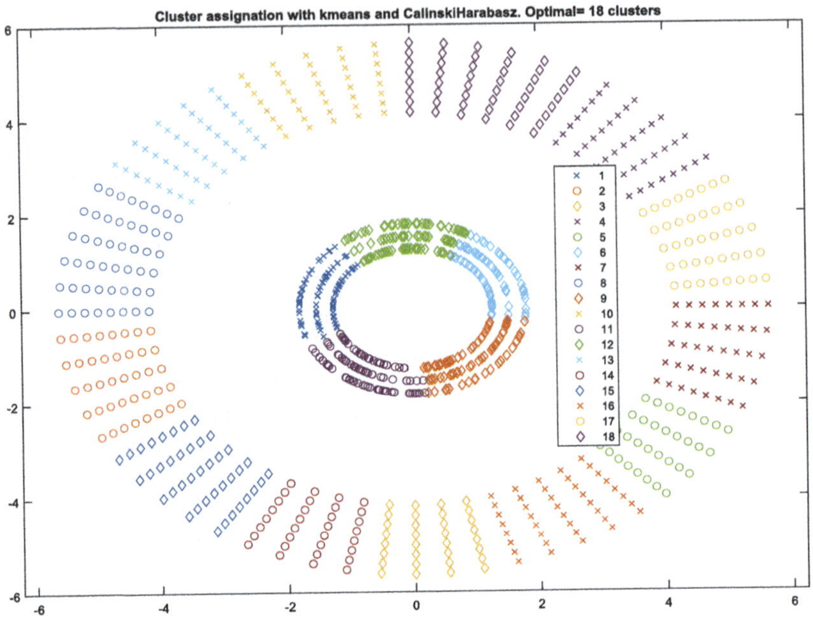

Fig. 3. Cluster in Cluster data - *k*-means algorithm - Calinski-Harabasz criterion.

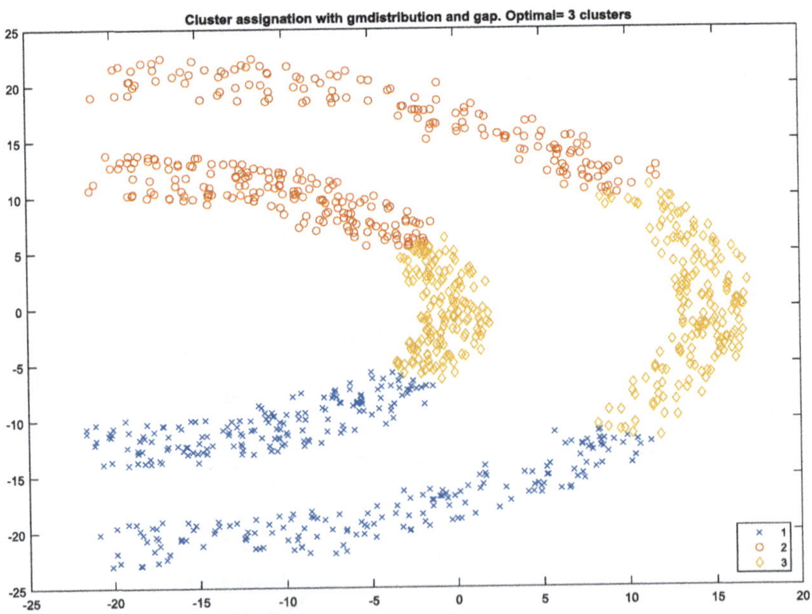

Fig. 4. Half Kernel data - Gaussian mixture distribution algorithm - Gap criterion.

The extension of this paper does not allow to include all the clustering figures, so, because of that, the results are shown in tables. There are six different tables, one for each original dataset, where it is possible to see if the clustering algorithm is able to cluster the data in a good way using different clustering evaluation criterion.

Table 1, for example, shows the number of clusters for each combination of clustering algorithm and evaluation criterion, and if the clustering obtained is Good or Wrong by using the Two Spirals dataset. In this case, as it is the most difficult dataset to cluster, all the obtained results are wrong.

Table 1. Two Spirals results

	Calinski-Harabasz		Davies-Bouldin		Gap criterion		Silhouette	
k-means	6	Wrong	4	Wrong	6	Wrong	4	Wrong
Agglomerative	6	Wrong	5	Wrong	6	Wrong	3	Wrong
Gaussian mixture distribution	6	Wrong	5	Wrong	6	Wrong	5	Wrong

Table 2 shows the results for Cluster in Cluster dataset. In this case, almost all the combinations obtain a good classification of the samples, but it is possible to emphasized the k-means and Silhouette, that creates only 6 clusters.

Table 2. Cluster in Cluster results

	Calinski-Harabasz		Davies-Bouldin		Gap criterion		Silhouette	
k-means	18	Good	10	Good	16	Good	6	Good
Agglomerative	31	Good	11	Good	10	Good	7	Good
Gaussian mixture distribution	17	Good	7	Good	3	Wrong	8	Good

Table 3 shows the results for Corners dataset. In this case, it is possible to remark that the Gaussian mixture distribution always classifies the samples in a Wrong way. Again, k-means and Silhouette is one of the combinations that uses less clusters to make a good classification.

Table 3. Corners results

	Calinski-Harabasz		Davies-Bouldin		Gap criterion		Silhouette	
k-means	12	Good	12	Good	11	Wrong	12	Good
Agglomerative	38	Good	12	Good	12	Good	14	Good
Gaussian mixture distribution	6	Wrong	4	Wrong	6	Wrong	6	Wrong

Table 4 shows the results for Half Kernel dataset. Table 5 shows the results for Crescent Full Moon dataset, and Table 6 shows the results for Outlier dataset.

Table 4. Half Kernel results

	Calinski-Harabasz		Davies-Bouldin		Gap criterion		Silhouette	
k-means	20	Good	6	Wrong	16	Good	6	Wrong
Agglomerative	32	Good	7	Wrong	13	Good	6	Wrong
Gaussian mixture distribution	6	Wrong	6	Wrong	3	Wrong	6	Wrong

Table 5. Crescent Full Moon results

	Calinski-Harabasz		Davies-Bouldin		Gap criterion		Silhouette	
k-means	13	Good	6	Good	8	Good	4	Good
Agglomerative	6	Good	6	Good	6	Good	5	Good
Gaussian mixture distribution	11	Good	6	Good	6	Good	4	Good

Table 6. Outlier results

	Calinski-Harabasz		Davies-Bouldin		Gap criterion		Silhouette	
k-means	16	Good	6	Good	8	Good	3	Wrong
Agglomerative	11	Good	4	Wrong	9	Good	3	Wrong
Gaussian mixture distribution	10	Good	4	Good	7	Good	3	Wrong

Table 7 shows a summary of the Good classification for each combination of algorithm and criterion. It is possible to see that there are three combinations that work good in 5 of the 6 dataset: k-means with Calinski-Harabasz, Agglomerative with Calinski-Harabasz and Agglomerative with Gap criterion.

It is important to remark that the original datasets used in this research are very complicated to divide in a good way using "simple" clustering algorithms. Instead of trying to find the real number of clusters (2 or 4 depending on the dataset), the number of clusters is increased and a classification that do not mix the original groups of data is searched. That is the reason why the proposed optimal number of clusters is higher that the original; but the most important fact is that the data is not mixed between groups.

Table 7. Summary of the results

	Calinski-Harabasz	Davies-Bouldin	Gap criterion	Silhouette
k-means	5	4	4	3
Agglomerative	5	3	5	3
Gaussian mixture distribution	3	3	2	2

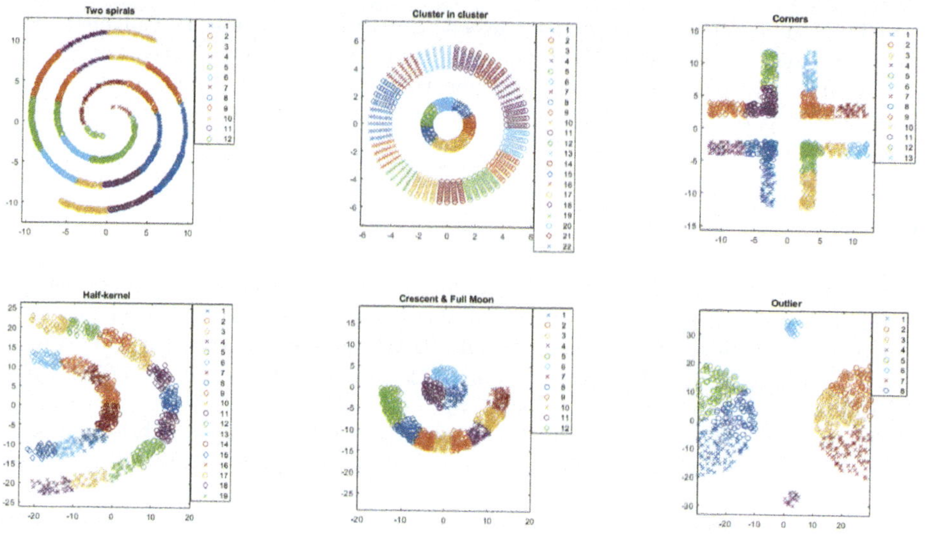

Fig. 5. Clustering created using k-means and Calinski-Harabasz.

4 Conclusions and Future Works

It is decided to finally implement k-means with Calinski-Harabasz as criterion to obtain the optimal number of clusters according to the easy way to assign new data to each cluster. Figure 5 shows the clusters created in the dataset using this combination of algorithm and criterion.

The main conclusion of this research is that a procedure that automatically selects the optimal number of clusters is developed. As future works, it will try to include this procedure to create a hybrid intelligent model without the necessity of manually choosing the number of local models. Previously, the authors developed a way to create hybrid intelligent models but, by now, the training process requires a lot of time to train, mainly because of the fact that several topologies for hybrid models are trained. With the use this research, it is desired to reduce this time and automate the model creation.

Acknowledgements. Álvaro Michelena's research was supported by the Spanish Ministry of Universities (https://www.universidades.gob.es/), under the "Formación de Profesorado Universitario" grant with reference FPU21/00932.

Antonio Díaz-Longueira's research was supported by the Xunta de Galicia (Regional Government of Galicia) through grants to Ph.D. (http://gain.xunta.gal), under the "Axudas á etapa predoutoral" grant with reference: $ED481A2023072$.

Míriam Timiraos's research was supported by the Xunta de Galicia (Regional Government of Galicia) through grants to industrial Ph.D. (http://gain.xunta.gal), under the Doutoramento Industrial 2022 grant with reference: $04\_IN606D\_2022\_2692965$.

Xunta de Galicia. Grants for the consolidation and structuring of competitive research units, GPC (ED431B 2023/49).

CITIC, as a center accredited for excellence within the Galician University System and a member of the CIGUS Network, receives subsidies from the Department of Education, Science, Universities, and Vocational Training of the Xunta de Galicia. Additionally, it is co-financed by the EU through the FEDER Galicia 2021-27 operational program (Ref. ED431G 2023/01).

References

1. Díaz-Longueira, A., et al.: Comparative analysis of intelligent techniques for categorization of the operational status of LifePo4 batteries. In: García Bringas, P., et al. (eds.) International Conference on Hybrid Artificial Intelligence Systems. LNCS, vol. 14001, pp. 545–553. Springer, Cham (2023). https://doi.org/10.1007/978-3-031-40725-3_46
2. García-Ordás, M.T., et al.: Clustering techniques selection for a hybrid regression model: a case study based on a solar thermal system. Cybern. Syst. **54**(3), 286–305 (2023)
3. Kools, J.: 6 functions for generating artificial datasets, MATLAB central file exchange (2023). https://www.mathworks.com/matlabcentral/fileexchange/41459-6-functions-for-generating-artificial-datasets. Accessed 31 May 2023
4. Michelena, Á., et al.: Hybrid classification model based on supervised techniques for denial of service attacks detection over COAP protocol. In: Jove, E., Zayas-Gato, F., Michelena, Á., Calvo-Rolle, J.L. (eds.) International Symposium on Distributed Computing and Artificial Intelligence, vol. 742, pp. 1–10. Springer, Cham (2023). https://doi.org/10.1007/978-3-031-38616-9_1
5. Michelena, Á., et al.: Machine learning based system for detecting battery state-of-health. In: García Bringas, P., et al. (eds.) International Conference on Soft Computing Models in Industrial and Environmental Applications, vol. 749, pp. 165–173. Springer, Cham (2023). https://doi.org/10.1007/978-3-031-42529-5_16

Wind Speed Virtual Sensor for Small Wind Turbine

Antonio Díaz-Longueira[1,2](✉), Paula Arcano-Bea[1], Míriam Timiraos[1,2,3], Álvaro Michelena[1], Francisco Javier de Cos Juez[1,4], and José Luis Calvo-Rolle[1,2]

[1] Ciencia y Técnica Cibernética (CTC), Department of Industrial Engineering, University of A Coruña, 15405 Ferrol, A Coruña, Spain
{paula.arcano,miriam.timiraos.diaz,alvaro.michelena,jose.rolle}@udc.es, fjcos@uniovi.es

[2] University of A Coruña, CITIC, Campus de Elviña, 15071 A Coruña, Spain
a.diazl@udc.es

[3] Fundación Instituto Tecnológico de Galicia, Department of Water Technologies, National Technological Center, 15003 A Coruña, Spain

[4] Instituto Universitario de Ciencias y Tecnologías Espaciales de Asturias (ICTEA), University of Oviedo, C. Independencia 13, 33004 Oviedo, Spain

Abstract. Renewable energy sources are increasing their importance and presence in the energy production sector due to their null or almost null greenhouse gas emissions. However, the main disadvantage of this type of system is the dependence on weather conditions. This research proposes using a virtual sensor to measure the wind speed 10 m above ground through variables measured at ground level. The modeling process followed a feature selection step before applying four machine learning techniques. The implemented virtual sensor accurately estimated the wind speed at 10 m, an interesting tool to eliminate the physical sensor's buy, installation, and maintenance.

Keywords: Green energy · Wind power · Virtual sensor

1 Introduction

In recent decades, the planet's energy needs have increased as the world population, industrial activities, transport, and our quality of life have grown. In 2021, the primary energy demand surged by 5.8%, surpassing the 2019 figures by 1.3% [4]. This increase is not a problem by itself, but the more use of fossil fuels to cover this needs. Fossil fuels are the main source of energy, an example of this circumstance is that, in 2022, the consumption of fossil fuels contributed to 82% of primary energy [4]. Fossil fuels provide an exceptional form of energy production, however, there are two principal disadvantages of this energy source: its availability and its emissions. Putting the focus on the second one, it is well-known that the negative impacts on the environment are caused by pollution and

F. Zayas-Gato et al. (Eds.): DCAI 2024, LNNS 1173, pp. 73–83, 2024.
https://doi.org/10.1007/978-3-031-73910-1_8

global warming. Conference Of the Parties (COP) highlighted the importance of phase-out the fossil fuels extraction and use in the COP28, the 28th United Nations Climate Change Conference. This action is essential in order to restrict global warming to 1.5 °C above pre-industrial levels, limiting the devastating effects that a greater warm could represent [10]. Organizations, corporations, and even governments must revert this situation through several lines of action [14]. With the intention of reducing greenhouse gas emissions, facilities that generate electricity through renewable energies are being promoted as a significant role in reducing or palliating the consequences of climate change. By that, renewable source energies have increased their relevance in society as a clean energy source with no or almost no greenhouse gas emissions. The European Green Pact represents an example of this trend signed in 2021 by EU countries, in which they commit to a zero-emission scenario by 2050 [6].

Photovoltaic and eolic are the most promising among the various renewable technologies. Both of them can be categorized based on the generation structure: large-scale installations and self-consumption installations. In eolic generation, the wind speed is an essential measure, as energy generated is directly proportional to the cube of the wind speed [9]. Result important monitor all the variables that condition the energy production, in order to estimate the energy production and make good energy management decisions. Different variables lead to different sensors with several technologies and implementations. High mounted sensors have a poor accessibility, as they are installed at turbine height, implying greater maintenance and revision time.

In [12] a digital twin-driven sensing methodology is used to develop a virtual sensor that estimates wind speed. On the other hand, in [11] the virtual wind speed sensor was developed with a data-driven approach, using historical wind farm data using data mining algorithms.

This paper deals with a real application of a self-consumption wind power installation in a bioclimatic house, developing a virtual sensor to measure the wind speed at 10 m, eliminating the need to mount a wind speed sensor 10 m above the ground, by using machine-learning methods. The composition of this document is organized as follows: Sect. 2 delineates the material and methods employed in conducting the investigative procedures. Section 3 presents the empirical trials conducted, alongside their consequent findings. Concluding this document, the final section synthesizes the inferences drawn and proposes trajectories for forthcoming inquiries.

2 Material and Methods

In this section, the material and methods used in the research will be detailed.

2.1 Material

A small wind turbine, and the data set obtained directly from this facility, are the materials used in the research.

Case of Study. In Xermade, Lugo, Galicia, Spain, it is located Sotavento, Experimental Wind Farm [1], where commercial, technical, educative, and formative activities take place. This facility has several wind turbines, showing the different technologies presented in the Galician wind sector. With a more educative focus, it has a demonstrative bioclimatic building designed with the climatology factors in mind, with the primary objective of promoting renewable energy saving and efficiency. This bioclimatic building has several sources that cover the electric, thermal, and refrigeration needs of the building. A wind turbine plays a fundamental role in accomplishing the building's electricity requirements. It is a BORNAY Inclin 1500, a two-blade turbine attached to a shaft that rotates a three-phase permanent magnet synchronous generator and housing that allows it to face the wind direction and maximize its efficiency. This turbine, with a nominal power of 1500 W at 12 m/s wind speed, is placed in an eight-meter tower. The start wind speed is 3.5 m/s, and at 14 m/s, an automatic brake is applied.

Dataset. The dataset is a private dataset, due to privacy aspects, and it is not accessible. It contains 25 features from 3 groups: meteorological, electrical, and temporal. The temporal group only contains one variable, registering the date and time of the sample. The electrical group contains six variables measured at the inverter side, while the remaining 18 are meteorological. However, as the main goal is to predict the wind speed at 10 m through meteorological features, the other features groups are not considered, resulting in a 18-variable dataset. Then, Table 1 summarizes the variables taken into account in the experiment setup. There are no structures or constructions that could affect the measurement of the sensors.

This paper primarily focuses on the wind speed at 10 m, disregarding other 10 m variables to simplify sensor installation and maintenance. The dataset, constructed with a 10-minute sample rate over a year, initially contained 52560 samples. After eliminating incorrect measures, it was refined to 50835 samples.

2.2 Methods

In this section, the methods used to indirect estimate the sensor measure in the research will be presented. Two main steps are followed to achieve proper prediction: first, the relationship of each variable with the wind speed at 10 m is calculated, and then regression techniques are applied.

Correlation Matrix. The correlation matrix, a statistical analysis, constitute one way for assessing the strength and direction of linear relationships between variables. It represents the pairwise correlations among variables in a symmetric square matrix [8]. This correlation values are shown by a number that ranges from −1 to 1. Values near zero mean no linear correlation. A positive strong correlation is represented by values near 1, while values near −1 indicate negative strong correlations.

Table 1. Meteorological variables descriptions, units and tags

Variable description	Units	Tag
Wind Speed	m/s	Speed
Wind direction	°	Direction
Amount of rain in 10 min	$l/m2$	Rain
Standard deviation of wind direction at 10 m	°	DirectionSD10m
Standard deviation of wind speed at 10 m	m/s	SpeedSD10m
Maximum wind direction at 10 m	°	BreakDirection10m
Wind direction at 10 m	°	Direction10m
Sunlight hours in 10 min	hour	SunlightHours
Relative humidity at 1.5 m	%	RH1.5 m
Atmospheric pressure	mBar	Pressure
Reduced atmospheric pressure	mBar	ReducedPreassure
Global solar radiation	W/m^2	GlobalSolarRadiation
Maximum wind speed at 10 m	m/s	Break10m
Dew Temperature	$^{\circ}C$	DewT1.5m
Ground temperature at −0.1 m	$^{\circ}C$	GroundT0.1m
Ambient temperature at 0.1 m	$^{\circ}C$	T0.1m
Ambient temperature at 1.5 m	$^{\circ}C$	T1.5m
Wind Speed at 10 m	m/s	Speed10m

Regression Techniques. The distinct regression techniques employed are presented below.

Recursive Least Squares Regressor. The Recursive Least Squares (RLS) algorithm is a method that minimizes the sum of squared errors, which are the differences between the predictions made by the model and the actual values [5]. The method uses a linear function to model the relationship between the target variable, y, and the features, $\mathbf{X}$, using the model parameters, $\mathbf{w}$. This relationship can be expressed as follows:

$$y(x, w) = w_0 + w_1 \times x_1 + w_2 \times x_2 + ... + w_n \times x_n \quad (1)$$

where n is the number of features. The hyperparameters affect how $\mathbf{w}$ is computed. The coefficients can be constrained to be positive, and the model can be configured to include or exclude the intercept term, w_0. If the intercept is not computed, it is set to 0.

Decision Tree Regressor. A Decision Tree (DT) method is a non-parametric method [3] that uses a tree diagram to split the data into smaller and more homogeneous groups at each node. The model predicts a numeric value when it reaches the final node. The model follows if-then conditions that are derived from rules learned based on the explanatory variables. The model's performance can be tuned by choosing the criteria for the best data split at each node and the maximum depth of the tree diagram.

Polynomial Regression. Polynomial regression (PN) is a type of linear regression that models the relationship between explanatory variables and the response variable using n-degree polynomials [13]. Similar to linear regression, the options include constraining the coefficients to be positive and fitting an intercept term or not. Additionally, the maximum degree of the polynomial is a hyperparameter.

K-Nearest Neighbours Regressor. K-Nearest Neighbors (KNN) is a non-linear algorithm [7] that predicts the value of an input data based on the mean of its K closest neighbors in the training data. Unlike the others algorithms presented, KNN does not train a model; it only stores the training data for later use. The algorithm can be tuned by changing the number of neighbors, K, and the weighting scheme for each neighbor.

3 Experiment Setup and Results

In this section, the experiment setup and their results are described.

3.1 Experiment Setup

The experiment setup is as follow, which includes the feature selection, techniques configuration and the metrics used to evaluate the performance.

Feature Selection. The feature selection is made based on the correlation matrix, selecting the four variables with stronger correlation to the target variablo.

Regression Techniques. Each regression technique's hyperparameters are adjusted to achieve the best prediction performance. The hyperparameters used are presented below.

Recursive Least Square. Four different models were tested, with the coefficients forced to be positive or not and the independent term calculated or not.

Decision Tree. Twenty different models were tested, using different values for the maximum diagram's depth, from one to ten, one by one. Two criteria based on the absolute and squared error were also tested.

Polynomial Regression. Sixteen models were tested, modifying the maximum polynomial degree and the independent term. All polynomial degrees, from second to ninth degree, were tested. The independent term was tuned to be calculated, and the bet was set to zero.

K-Nearest Neighbors. Twenty different models were used, testing all combinations from one to ten neighbors and two different functions to establish the weight of each neighbor in the prediction: a uniform function, where all neighbors have the same weight, and a distance function, where the weight of the neighbor is inversely proportional to the distance to the query point.

Model Evaluation. Several metrics were used to evaluate the performance of the models. These metrics are the Mean Squared Error (MSE), Mean Absolute Error (MAE), Symmetric Mean Absolute Percentage Error (SMAPE), and Coefficient of Determination (R^2), defined as [2]. Metrics are evaluated in two stages. Initially, a 10 k-fold cross-validation is conducted, using 80% of the samples to determine the optimal model configuration. Once each technique is tuned, the remaining 20% of the samples are used for model validation.

3.2 Results

In this section, the results obtained during the experiments are shown.

Feature Selection. Figure 1 shows the correlation between the different meteorological variables. Special attention is paid to Speed10m, ignoring SpeedSD10m, DirectionSD10m, Direction10m, and Break10m, as they come from measures taken at 10 m high. This Figure establish that the wind speed at 10 m (Speed10m) dramatically correlates with the wind speed at ground level, with a correlation value of 0.79. The following variables with the highest correlation are the atmospheric pressure (Pressure), reduced pressure (ReducedPreassure), and the ground temperature at 0.1 m (GroundT0.1 m), with a correlation value of −0.33, −0.31 and −0.2, respectively. After selecting the most correlated variables to indirectly predict the 10 m wind speed sensor's output, Fig. 2 displays the virtual sensor topology, using four ground mounted sensors as input.

Regression Techniques Results. Table 2 shows each model's configuration selected to get the best prediction possible. RLS best configuration is obtained after calculating the independent term and not forcing positive coefficients, while the KNN technique achieved its best performance using 3 neighbors and a distance-based weight function. DT's best models uses an absolute error criteria and a maximum's diagram depth of 15 splits. In the case of PN regression, a 4 degree equation without independent term obtained the best behavior.

Table 3 displays the average metrics values obtained in the cross validation train phase by this latter models. In bold are highlighted the best results of each metric, always obtained by the KNN model.

Figures 3, 4, 5 and 6 display a visual representation of the metrics' values achieved in the cross-validation. The boxplots shows, in every metric, a no significant deviation in the value obtained along all the different folds in the training phase. As Table 3 represents, KNN raises as the best forecasting technique.

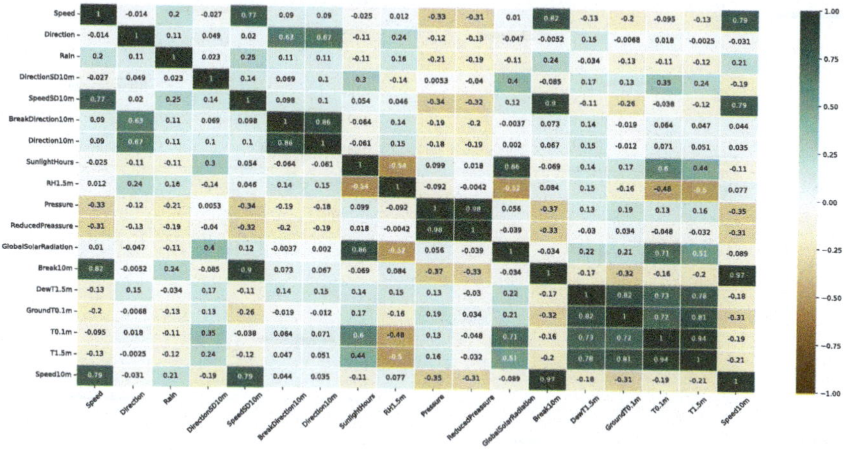

Fig. 1. Correlation Matrix between meteorological variables

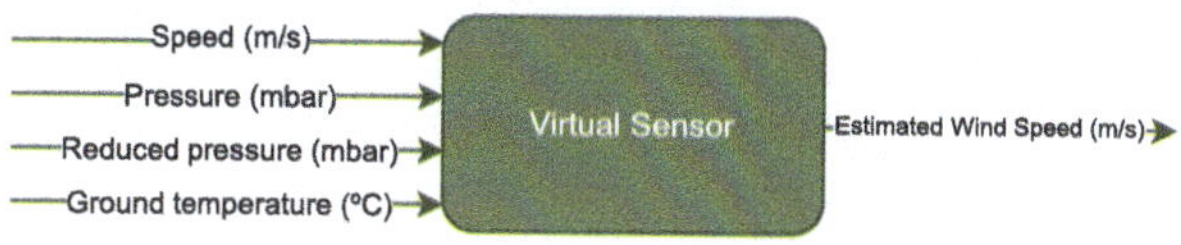

Fig. 2. Virtual sensor topology after Correlation Matrix analysis

Table 2. Model hyperparameters with the best performance

Technique	Hyperparameter	Value
RLS	Independent term	Calculated
	Positive coefficient	Not forced
DT	Maximum diagram depth	15
	Criterion	Absolute error
PN	Maximum degree	4
	Independent term	Not calculated
KNN	Number of neighbors	3
	Weight function	Distance

Table 3. Average metric values during the models training phase

Technique	MSE (m^2/s^2)	MAE (m/s)	SMAPE (%)	R^2
RLS	2.48	1.23	29.12	0.65
DT	1.77	0.94	59.98	0.75
PR	2.13	1.13	28.02	0.70
KNN	**0.85**	**0.63**	**16.22**	**0.88**

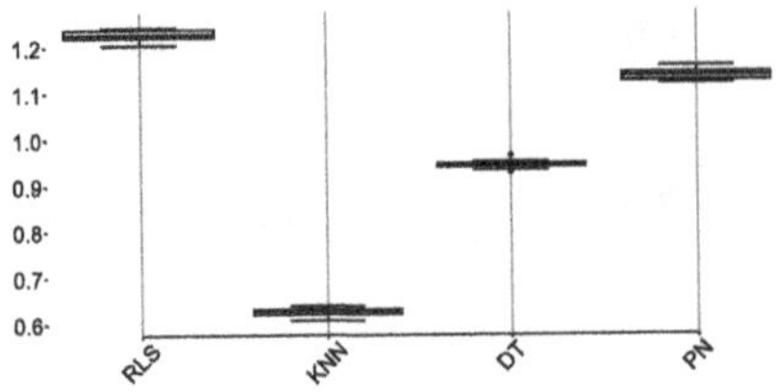

Fig. 3. MAE boxplot

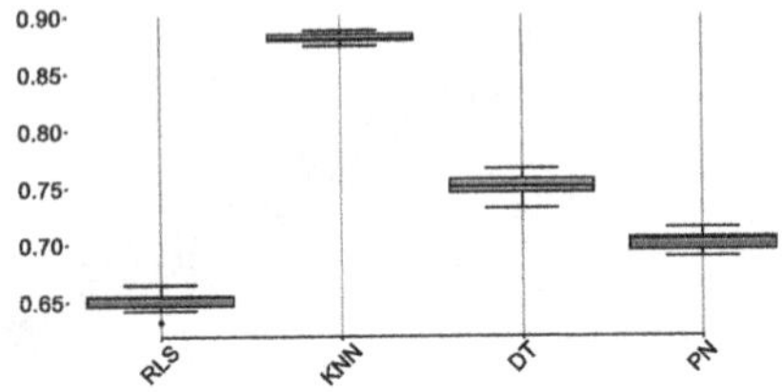

Fig. 4. R^2 boxplot

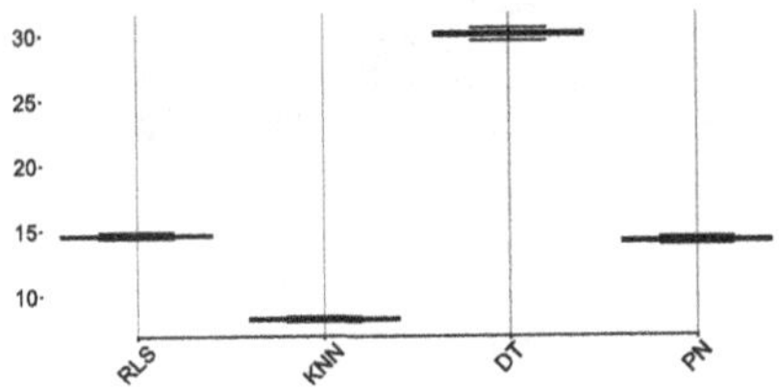

Fig. 5. SMAPE boxplot

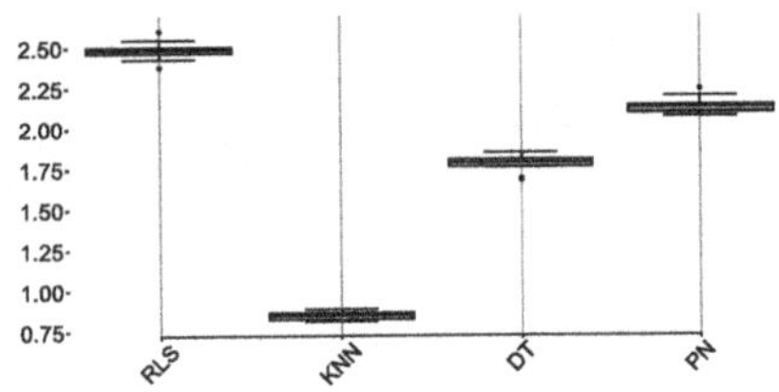

Fig. 6. MSE boxplot

Table 4 shows the MSE, MAE, SMAPE and R^2 values obtained during the validation phase. In bold are highlighted the best results of each metric. The values obtained in this validation phase are almost identical to the ones obtained in the training phase, meaning that the models are not overfitted and the prediction is adequate.

Table 4. Metric values during the models validation phase

Technique	MSE (m^2/s^2)	MAE (m/s)	SMAPE (%)	R^2
RLS	2.42	1.22	29.04	0.65
DT	1.68	0.92	23.54	0.76
PR	2.06	1.12	27.88	0.70
KNN	**0.79**	**0.60**	**15.82**	**0.89**

Virtual Sensor Evaluation. Figure 7 shows the values predicted by the KNN model in the validation phase. The predicted values are displayed versus the actual values. A proper forecasting is achieved, with all predicted values close to the optimal diagonal dotted line.

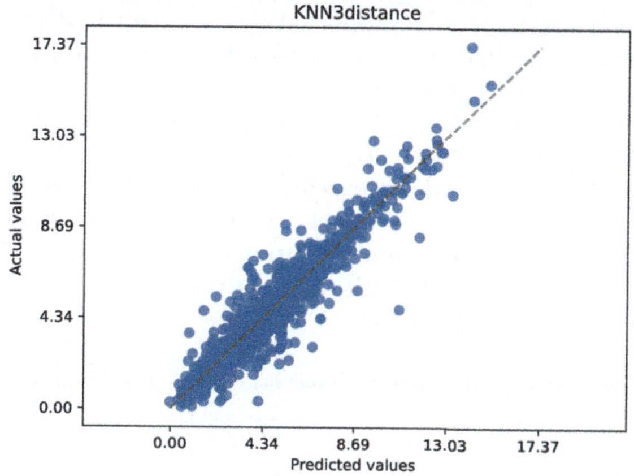

Fig. 7. Validation KNN predicted values

4 Conclusion and Future Works

This research presents a virtual sensor that measures the wind speed 10 m above the ground from measurements taken from sensors at ground level, eliminating the physical sensor mounted 10 m high. Several regression techniques, linear and non-linear, were tested to find the best prediction.

This virtual sensor, based on K-Nearest Neighbor with 3 neighbors and distance-based weights, uses wind speed at ground level, temperature, atmospheric pressure, and atmospheric reduced pressure, obtaining a Mean Absolute Error of 0.60 m/s and a Mean Squared Error of 0.79 m^2/s^2. It achieves a Symmetric Mean Absolute Percentage Error of 7.91% and a Coefficient of Determination of 0.89. The performance of other techniques is worse in every metric. These results set the virtual sensor as an effective method for validating the measurements obtained from a physical sensor, thus increasing the reliability of the prediction process, eliminating the need of a recursive sensor, and translating in maintenance savings. Also, anomaly detection could be made based in deviations between the virtual and physical sensor.

In future works, to eliminate the need for 10-meter sensors, other virtual sensors can be developed in 10-meter variables that a virtual sensor can measure. Additionally, to improve the prediction, other techniques can be applied.

Acknowledgments. Antonio Díaz-Longueira's research was supported by the Xunta de Galicia (Regional Government of Galicia) through grants to Ph.D. (http://gain.xunta.gal), under the "Axudas á etapa predoutoral" grant with reference: $ED481A$-2023072.

Míriam Timiraos's research was supported by the Xunta de Galicia (Regional Government of Galicia) through grants to industrial Ph.D. (http://gain.xunta.gal), under the Doutoramento Industrial 2022 grant with reference: $04\_IN606D\_2022\_2692965$

This work has been supported by Centro Mixto de Investigación UDC-NAVANTIA (IN853C 2022/01), funded by GAIN (Xunta de Galicia) and ERDF Galicia 2021–2027.

Álvaro Michelena's research was supported by the Spanish Ministry of Universities (https://www.universidades.gob.es/), under the "Formación de Profesorado Universitario" grant with reference FPU21/00932.

Xunta de Galicia. Grants for the consolidation and structuring of competitive research units, GPC (ED431B 2023/49)

CITIC, as a center accredited for excellence within the Galician University System and a member of the CIGUS Network, receives subsidies from the Department of Education, Science, Universities, and Vocational Training of the Xunta de Galicia. Additionally, it is co-financed by the EU through the FEDER Galicia 2021-27 operational program (Ref. ED431G 2023/01).

Disclosure of Interests. The authors have no competing interests.

References

1. Sotavento web page (2023). https://www.sotaventogalicia.com. Accessed 4 Mar 2024
2. Botchkarev, A.: Performance metrics (error measures) in machine learning regression, forecasting and prognostics: properties and typology. arXiv preprint arXiv:1809.03006 (2018)
3. Czajkowski, M., Kretowski, M.: The role of decision tree representation in regression problems - an evolutionary perspective. Appl. Soft Comput. **48**, 458–475 (2016). https://doi.org/10.1016/j.asoc.2016.07.007, https://www.sciencedirect.com/science/article/pii/S1568494616303325
4. Energy Institute Statistical Review of World Energy: Energy Institute Statistical Review of World Energy 2023 (2023)
5. Engel, Y., Mannor, S., Meir, R.: The kernel recursive least-squares algorithm. IEEE Trans. Signal Process. **52**(8), 2275–2285 (2004)
6. Haller, A.: Influence of agricultural chains on the carbon footprint in the context of European green pact and crises. Agriculture **12**(6), 751 (2022)
7. Imandoust, S.B., Bolandraftar, M., et al.: Application of k-nearest neighbor (KNN) approach for predicting economic events: theoretical background. Int. J. Eng. Res. Appl. **3**(5), 605–610 (2013)
8. Ivanov, A., Bezyayev, A., Gazin, A.: Simplification of statistical description of quantum entanglement of multidimensional biometric data using symmetrization of paired correlation matrices. J. Comput. Eng. Math. **4**, 3–13 (2017). https://doi.org/10.14529/jcem170201
9. Karakislak, I., Hildebrand, J., Schweizer-Ries, P.: Exploring the interaction between social norms and perceived justice of wind energy projects: a qualitative analysis. J. Environ. Policy Plann. **25**(2), 155–168 (2023)
10. Khan, M.A., Khan, M.Z., Zaman, K., Naz, L.: Global estimates of energy consumption and greenhouse gas emissions. Renew. Sustain. Energy Rev. **29**, 336–344 (2014)
11. Kusiak, A., Zheng, H., Zhang, Z.: Virtual wind speed sensor for wind turbines. J. Energy Eng. **137**(2), 59–69 (2011). https://doi.org/10.1061/(ASCE)EY.1943-7897.0000035, https://ascelibrary.org/doi/abs/10.1061/(ASCE)EY.1943-7897.0000035
12. Li, Y., Shen, X.: A novel wind speed-sensing methodology for wind turbines based on digital twin technology. IEEE Trans. Instrum. Measur. **71**, 1–13 (2022). https://doi.org/10.1109/TIM.2021.3139698

13. Ostertagová, E.: Modelling using polynomial regression. Procedia Eng. **48**, 500–506 (2012)
14. Yin, Z., et al.: Implication of electrification and power decarbonization in low-carbon transition pathways for China, the US and the EU. Renew. Sustain. Energy Rev. **183**, 113493 (2023)

Reconstructing Turbulence-Distorted Wavefronts Through Laser-Beam Profiles

Alejandro Buendía-Roca[1(✉)], Saúl Pérez-Fernández[1], Carlos González-Gutiérrez[1,2], Santiago Iglesias-Álvarez[1], Javier Rodríguez-Rodríguez[1], Julia María Fernández-Díaz[1], Fernando Sánchez-Lasheras[1,3], Nazim Ali Bharmal[4], José Luis Calvo-Rolle[5], and Francisco Javier de Cos Juez[1,6]

[1] Institute of Space Sciences and Technologies of Asturias (ICTEA), Oviedo, Spain
buendiaalejandro@uniovi.es

[2] Computer Sciences Department, University of Oviedo, Campus de Viesques, 33024 Gijón, Spain

[3] Department of Applied Mathematics, University of Oviedo, C/Leopoldo Calvo Sotelo, 18, 33007 Oviedo, Spain

[4] Centre for Advanced Instrumentation, Department of Physics, Durham University, South Road, Durham DH1 3LE, UK

[5] Department of Industrial Engineering, University of A Coruña, CTC, Avda. 19 de Febrero s/n, 15405 Ferrol, A Coruña, Spain

[6] Department of Exploitation and Prospection of Mines, University of Oviedo, C/Independencia 13, 33004 Oviedo, Spain

Abstract. Adaptive optics (AO) compile a variety of techniques that produce high-quality images in ground-based telescopes. Most commonly used AO techniques require the use of laser guide stars, which introduce errors due to the limited width and low altitude of these artificial stars, resulting in issues such as anisoplanatism and cone-effect problems. This study proposes an alternative AO system that aims to eliminate these issues. In contrast with similar techniques such as the Projected Pupil Plane Pattern (PPPP), that uses two pictures, Wavefronts Obtained from Measurements from Beam-profiles through Atmospheric Turbulence (WOMBAT) requires only one picture, reducing the complexity of the system. Even with only one image, WOMBAT manages to achieve similar results to PPPP in wavefront correction, offering potential advancements in ground-based telescope imaging quality.

Keywords: Adaptive Optics · Astronomical instrumentation · Machine Learning

1 Introduction

Due to temperature differences in small scales throughout the atmosphere, light coming to ground-based telescopes does not travel in a straight line. These temperature differences can vary between consecutive points producing variations

F. Zayas-Gato et al. (Eds.): DCAI 2024, LNNS 1173, pp. 84–93, 2024.
https://doi.org/10.1007/978-3-031-73910-1_9

in the air refractive index, what is translated to blurry images in telescopes. As ground-base telescopes are increasing their size, such as Extremely Large Telescope (ELT) [6], the presence of atmosphere adds errors and uncertainties that should be carefully taken into account in the study of celestial bodies.

Adaptive Optics (AO) techniques take into account the disturbances in the wavefront of incoming light and try to correct them before reaching the telescope [11]. However, these techniques are not exempt from challenges, particularly AO systems usually consist on a variety of optical components that measure how distorted a wavefront is and calibrates to produce the opposite perturbance, resulting in the recovery of the original (non-distorted) light. Nevertheless, these techniques are not exempt of problems, as it can be a hard task to do due to the randomness of atmospheric behaviour.

All these problems grow for larger telescopes, as the pupil surface increases. Optical components become larger and more expensive, leading to rises in the budgets and more complex structures. Projected Pupil Plane Pattern (PPPP) [1] proposes an alternative, replacing some of the components with a collimated laser and a camera and a computer, to implement the use of Artificial Neural Networks (ANNs) [16]. As its successor, Wavefronts Obtained from Measurements from Beam-profiles through Atmospheric Turbulence (WOMBAT) system aims to improve PPPP results while reducing even more its requirements.

In this work, we present the results of testing of WOMBAT over different turbulent conditions and compare it with PPPP results. In Sect. 2, the main concepts of AO and WOMBAT are explained; in Sect. 3, ANNs are described; Sect. 4 summarizes the datasets used for the experiment and presents our results, comparing them with PPPP.

2 Adaptive Optics

AO systems aim to mitigate atmospheric effects on light reaching ground-based telescopes. They measure the distortion produced by the atmosphere in the light coming from celestial sources and, through a process of reconstruction, they restore the shape of the wavefront as it came prior to traversing atmospheric turbulences.

The main components that form an AO system are the following [5]: a wavefront sensor, usually a Shack-Hartmann wavefront sensor (SH-WFS), an array of lenses that measure the local curvature of the light reaching the telescope; a deformable mirror (DM), a flexible reflective surface placed over an array of actuators that adjust the mirror's shape based on information of the SH-WFS; and a reconstructor, a system designed to translate the measurements of the SH-WFS into values for the actuators to curve the DM.

Traditional AO systems use guide stars as references to characterise the turbulent atmosphere over the telescopes. There are two types of guide stars: Natural Guide Stars (NGS), which are well-known stars that are close to the target of study, and Laser Guide Stars (LGS), lasers that excite certain parts of the atmosphere creating an artificial star with known properties. However, there are limitations to their use. NGSs are not a reliable for every study, as they may be too far from the target, what is even more complicated as telescopes and their resolution grow. On the other hand, LGSs present problems, such as cone-effects. As the excitation of atmosphere occur at certain layers (20 km for Rayleight scattering or 90 km for sodium guide stars) [14], they do not recover the whole turbulence, instead seen turbulence takes the form of a cone between the LGS and the WFS.

Focusing on LGSs, there have been several attempts to improve their performance, like Laser Tomography AO (LTAO) [12], that use several LGSs around the target to reconstruct the turbulence, or Projected Pupil Plane Pattern (PPPP), which uses a collimated laser through the pupil of the telescope to avoid the cone effect.

WOMBAT system aims to improve PPPP system [2]. Both system use cameras taking pictures of a laser which gathers the information of the different turbulent layers. Nevertheless, perturbations in the laser beam are not obvious enough to be directly reconverted into useful information, but here is where Artificial Neural Networks excel. Changes in the laser follow a non-linear relation that can be learnt by ANNs. In particular, Convolutional Neural Networks (CNNs) have demonstrated to improve significantly the performance of PPPP [16].

WOMBAT proposes the emission of two lasers from the telescope pupil, one collimated and one diverging (Fig. 1). The addition of the divergent laser breaks the magnitude-distance degeneracy present in PPPP and lowers the needs of two laser pictures to just one, leading to a decrease in the computational requirements of the ANNs.

The goal of both PPPP and WOMBAT systems is to recover the shape of the turbulence layers through which light travels. To achieve this, turbulences are represented using Zernike polynomials. Zernike polynomials are a system of orthogonal polynomials over the unit circle [8]. Nevertheless, this decomposition adds a source of error, as it is not possible to work with an infinite sequence of coefficients. As a sufficiently good enough approximation, we choose to work with 55 Zernike polynomials (Fig. 2), ignoring the first term, which is the mean value and its corrected in a separated mirror. The choice of this number of Zernike polynomials is based on the resolution of the DM used in the experiments performed for PPPP [15].

3 Neural Networks

Artificial neural networks simulate the connections of neurons in the brain to recognise the hidden patterns in datasets [9]. Supervised machine learning, a process where artificial neural networks are trained using examples of desired outputs corresponding to given inputs, allows ANNs to identify relevant features and learn to calculate the intended output.

ANNs have shown to be useful in many different fields performing different tasks. In astronomical applications, they have been used in the study of Cosmic

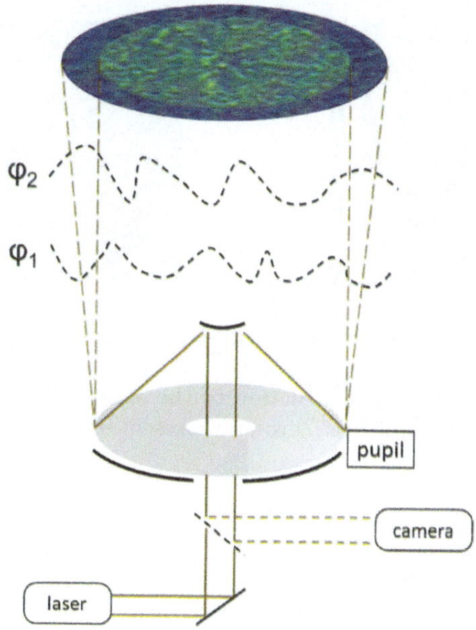

Fig. 1. Scheme of WOMBAT system. Light from a laser goes through the telescope pupil and a defocusing lens, recreating the inverse path light follows. As it goes up, the laser compiles the information of the turbulent layers. This information is captured by a camera and processed by a computer.

Microwave Background [3], in the detection of exoplanets [7] or for predictions in the time evolution of atmospheric turbulences [10].

Given that WOMBAT requires the use of images to reconstruct turbulences, Convolutional Neural Networks (CNNs) are chosen as a suitable candidate to retrieve the necessary data. These networks apply convolutional filters to the input data, aiming to uncover hidden patterns. These filters capture the information contained in the images, capturing the spatial correlation between pixels.

In our case, due to the spatial relations between consecutive points and their closely related optical paths, CNNs are able to recognise the differences and to recover the shape of the atmospheric turbulences. The structure chosen for WOMBAT is formed by three convolutional cells, formed by a convolutional layer, PReLU as activation function and a pooling layer, ending in a fully connected layer with 55 outputs. Learning rate exponentially decays with a starting value of 0.005 lowering to 0.0005 in the 200 training epochs. This structure is summarized in Fig. 3.

4 Experiments

Data used for the experiments comes from simulations replicating the path of the laser through the telescope and up to high layers of the atmosphere, where the

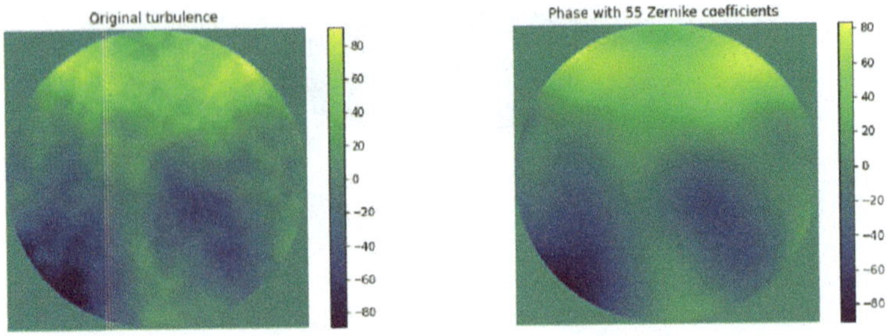

Fig. 2. Example of 10 turbulence layers with equal relative strength collapsed as seen from the telescope pupil. Left picture shows the turbulence as seen by the telescope and right picture shows the reconstruction of the same screen using only 55 Zernike coefficients. The coefficients of this Zernike representation conform the output of the ANN.

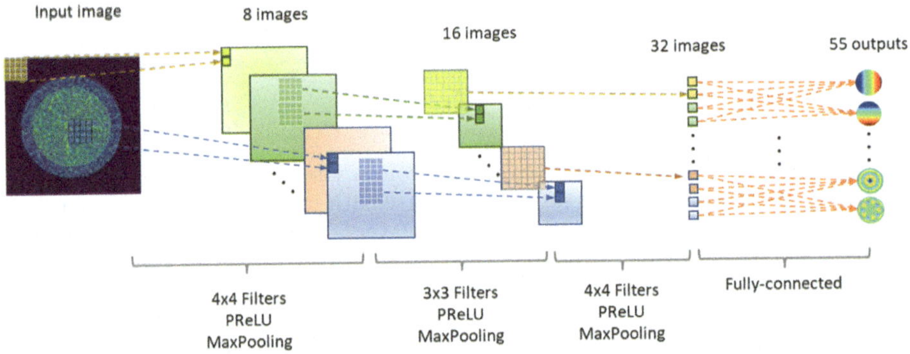

Fig. 3. Neural network structure used along this study. Input images are 64×64 or 128×128 pixels and 55 Zernike coefficients are extracted from the collapsed turbulence as output. The training lasts 200 epochs and the learning rate exponentially decays from 0.005 to 0.0005 for last epoch.

laser is photographed on-axis. Simulations emulate the propagation of the laser through random turbulence screens along its path [4], placed at random heights between 1 and 15 km and with different values of coherent lengths (r_0), 8, 12 and 16 cm@500 nm, which represents the length in which the wavefront maintains coherence, i.e. a consistent phase relationship. These values are wavelength-dependent (they are given for 500 nm wavelength as reference) and represent from poor to good seeing conditions. Upon reaching the photograph point, Zernike coefficients describing the collapsed turbulence are saved. An example of the final images is shown in Fig. 4.

Simulations assume that there is always a turbulence layer at ground level and that all layers have the same relative strength. Also, as said in Sect. 2, there is always an intrinsic error due to the finite number of Zernike coefficients used to recreate the atmosphere.

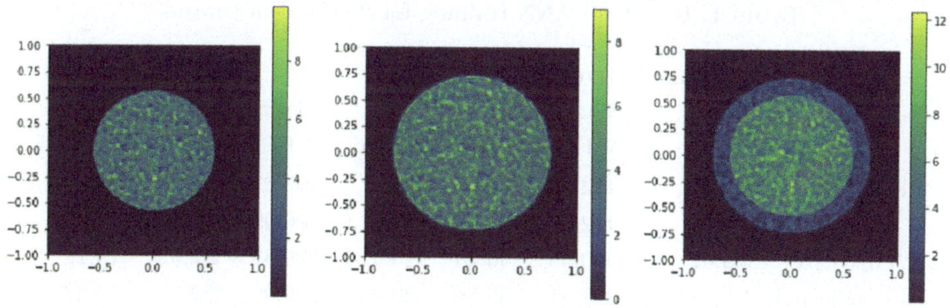

Fig. 4. Pictures of the laser as propagated through the atmosphere shown in Fig. 2. Left picture shows the collimated beam, middle picture shows the defocused beam and right picture represents the final image collected by WOMBAT camera.

Datasets used for experiments are generated from simulations with 2, 5 or 10 turbulent screens, fixing one at floor level and setting the rest at random heights between 1 and 15 km, all of them having $r_0 = 12\,\mathrm{cm}@500\,\mathrm{nm}$. WOMBAT pictures are saved with a resolution of 128×128 pixels and downsized to 64×64 pixels. For each number of turbulence layers, 500,000 samples have been simulated, keeping 50,000 of them as test. For different values of r_0 (8 and 16 cm@500 nm), only a test sample of 50,000 samples has been simulated.

The idea behind this distribution is to study WOMBAT's performance with a good but not optimal atmospheric conditions and to check its behaviour with better and worse conditions. Regarding the size of the images, we compare the ability of the system and networks to discern details by increasing or decreasing the size of the inputs.

How the networks fit the data is measured by the wavefront error (WFE). This error represents how distant is the wavefront from a plain wavefront. This value can be expressed by the following equation [8]:

$$WFE = \left(\frac{1}{A}\iint (\Phi(x,y) - \overline{\Phi})^2 dA\right)^{1/2} \simeq \left(\sum_{i=2}^{N_{zern}} z_i^2\right)^{1/2} \tag{1}$$

where Φ is the turbulent phase, $\overline{\Phi}$ its mean value and z_i is the value of the coefficient for the i-th Zernike polynomial in Noll's notation [8].

The first experiment is to train our ANN with images of 2, 5 and 10 turbulence layers and to test the training with the dataset with same number of layers with all values of r_0. The results of these tests can be seen in Tables 1, 2 and 3. In them, $WFE_{original}$ refers to the mean value of WFE that is measured from the simulated turbulences, with 55 Zernike coefficients; WFE_{pred} is the mean of the WFE predicted by our ANN architecture; WFE_{res} shows the mean WFE of the differences between the predicted and the original wavefront, called residual WFE; and $Error\%$ is the percentage of the WFE that is not corrected, referred to the original wavefront.

Table 1. Results of ANN training for 2 turbulent screens.

r_0	size	2 turbulence screens						
		$WFE_{original}$	WOMBAT			PPPP		
			WFE_{pred}	WFE_{res}	Error %	WFE_{pred}	WFE_{res}	Error %
8 cm	64 pxls	426.28	412.33	81.01	20.81	409.61	76.38	19.58
	128 pxls		412.52	80.93	20.81	406.99	77.02	19.77
12 cm	64 pxls	305.62	300.58	56.24	20.41	299.88	52.62	19.12
	128 pxls		297.81	55.55	20.26	299.99	52.75	19.23
16 cm	64 pxls	240.48	238.27	45.12	20.55	238.52	42.32	19.27
	128 pxls		234.79	44.42	20.19	238.79	42.51	19.36

Table 2. Results of ANN training for 5 turbulent screens.

r_0	size	5 turbulence screens						
		$WFE_{original}$	WOMBAT			PPPP		
			WFE_{pred}	WFE_{res}	Error %	WFE_{pred}	WFE_{res}	Error %
8 cm	64 pxls	108.45	103.89	23.22	23.49	103.85	22.09	22.34
	128 pxls		104.30	23.37	23.68	103.99	22.18	22.44
12 cm	64 pxls	77.46	74.90	16.16	23.14	75.63	15.62	22.43
	128 pxls		75.74	16.09	23.34	75.52	15.43	22.20
16 cm	64 pxls	60.56	58.91	12.82	23.32	59.73	12.57	22.93
	128 pxls		59.69	12.93	23.56	59.69	12.35	22.50

We achieve up to a 80% correction for all the turbulent conditions and for both image sizes, confirming that WOMBAT systems works under simulations.

Even if PPPP still shows better results that WOMBAT, the differences are slight, there is no more than 1.2% improvement with PPPP. Same happens with different image sizes, there is no meaningful improvement in choosing bigger, more detailed images, as ANNs work similarly. These results mean that WOMBAT can work with less computationally demanding data, what can be translated to a simplification of the system.

Nevertheless, the idea behind these experiments is to check the behaviour of the system when atmospheric conditions for training and testing are mixed, to generalize its training. Another approach was to perform Transfer Learning (TL) [13] on the different datasets. By training with a dataset and re-using those weights, we can improve the performance of the system, as ANN sees all datasets. Training for this experiment started with 200 epochs with the 2 layers dataset and 50 epoch for 5 and 10 layers datasets, all of them with the same ANN architecture and decaying learning rate.

The results with TL are promising: with 64 × 64 images, we manage to achieve 19.74%, 20.85% and 15.85% correction with 2 layers for r_0 values of 12, 16 and

Table 3. Results of ANN training for 10 turbulent screens.

		10 turbulence screens						
r_0	size	$\text{WFE}_{original}$	WOMBAT			PPPP		
			WFE_{pred}	WFE_{res}	Error %	WFE_{pred}	WFE_{res}	Error %
8 cm	64 pxls	38.44	37.29	8.16	23.43	37.43	7.87	22.59
	128 pxls		37.46	8.08	23.26	36.97	7.69	22.07
12 cm	64 pxls	27.41	26.89	5.67	23.09	26.90	5.48	22.30
	128 pxls		27.03	5.56	22.82	26.86	5.34	21.88
16 cm	64 pxls	21.57	21.26	4.57	23.44	21.25	4.43	22.73
	128 pxls		21.36	4.47	22.95	21.27	4.29	22.03

8 cm respectively; 48.42%, 48.45% and 45.86% with 5 layers and 77.29%, 75.48% and 77.55% for 10 layers. For 128 × 128 images, the results are similar: 22.72%, 22.70% and 22.56% for 2 layers; 49.24%, 48.16% and 48.05% for 5 layers and 77.19%, 76.73% and 76.78% for 10 layers, each for r_0 values of 12, 16 and 8 cm respectively.

In order to make a comparison with a dataset different from those of training, we simulated a dataset with 8 turbulence layers. The results of evaluating this training with the new dataset are shown in Table 4.

Table 4. Results of Transfer Learning for a dataset with 8 turbulence layers.

		8 turbulence screens			
r_0	size	$\text{WFE}_{original}$	WOMBAT		
			WFE_{pred}	WFE_{res}	Error %
8 cm	64 pxls	53.51	41.47	15.68	30.43
	128 pxls		42.48	14.95	29.25
12 cm	64 pxls	38.13	29.77	11.01	30.04
	128 pxls		30.15	10.60	28.96
16 cm	64 pxls	29.99	23.45	8.72	30.29
	128 pxls		23.67	8.39	29.16

5 Conclusions and Future Lines

WOMBAT has demonstrated to be a useful AO system under simulations. These results resume a variety of conditions, from simpler turbulences to more complex ones formed by more layers or with lower coherent lengths. We manage to get up to 80% correction in the incoming wavefronts and we built up a model that can recover the shapes of the turbulences for a variety of atmospheric conditions. Comparing with PPPP, we achieve similar results, with only around a 1% correction difference, despite using just one image. Even though WOMBAT is behind PPPP that small amount, every simulation has an on-axis set up, where PPPP is supposed to work better.

As future lines of investigation, off-axis simulations will be studied. PPPP is designed to work on-axis, following the path of collimated light. On the other hand, the defocused beam used on WOMBAT will allow to recover a wider area, what can lead to multi-object observation. Another following step for WOMBAT is the implementation on an optical bench, where real tests can be done. Aiming to implement WOMBAT in a real telescope in the future, testing on a optical bench will allow to test the functionality of the system over real data and to do necessary adjustments. The implementation in a real telescope will be a challenge due to the nature of WOMBAT system. Given the need of a wide laser, acquiring one with enough power is a difficult task, even though [16] reduced the power required to ∼200 W. The use of both collimated and defocusing laserbeams has to be properly assembled to the telescope system.

References

1. Bharmal, N.A., Myers, R.M., Yang, H.: PPPP: an on-sky experiment for zero-cone effect LGS alternative. In: Close, L.M., Schreiber, L., Schmidt, D. (eds.) Adaptive Optics Systems VI. Society of Photo-Optical Instrumentation Engineers (SPIE) Conference Series, vol. 10703, p. 1070362, July 2018. https://doi.org/10.1117/12.2313166
2. Buscher, D.F., Love, G.D., Myers, R.M.: Laser beacon wave-front sensing without focal anisoplanatism. Opt. Lett. **27**(3), 149–151 (2002). https://doi.org/10.1364/OL.27.000149
3. Casas, J.M., et al.: Multi-frequency point source detection with fully convolutional networks: performance in realistic microwave sky simulations. Astron. Astrophys. **658**, A110 (2022). https://doi.org/10.1051/0004-6361/202141874
4. Chan, S.H., Chimitt, N.: Computational imaging through atmospheric turbulence. Found. Trends® Comput. Graph. Vis. **15**(4), 253–508 (2023). https://doi.org/10.1561/0600000103
5. Davies, R., Kasper, M.: Adaptive optics for astronomy. Annu. Rev. Astron. Astrophys. **50**, 305–351 (2012). https://doi.org/10.1146/annurev-astro-081811-125447
6. Hippler, S., et al.: Single conjugate adaptive optics for the ELT instrument METIS. Exp. Astron. **47**, 65–105 (2019). https://doi.org/10.1007/s10686-018-9609-y
7. Iglesias Álvarez, S., et al.: One-dimensional convolutional neural networks for detecting transiting exoplanets. Axioms **12**(4), 348 (2023). https://doi.org/10.3390/axioms12040348
8. Lakshminarayanan, V., Fleck, A.: Zernike polynomials: a guide. J. Mod. Opt. **58**, 1678–1678 (2011). https://doi.org/10.1080/09500340.2011.633763
9. LeCun, Y., Bengio, Y., Hinton, G.: Deep learning. Nature **521**, 436–444 (2015). https://doi.org/10.1038/nature14539
10. Pérez, S., et al.: Enhancing open-loop wavefront prediction in adaptive optics through 2D-LSTM neural network implementation. Photonics **11**(3), 240 (2024). https://doi.org/10.3390/photonics11030240

11. Roddier, F.: Adaptive Optics in Astronomy. Cambridge University Press, Cambridge (1999)
12. Tatulli, E., Ramaprakash, A.N.: Laser tomography adaptive optics: a performance study. J. Opt. Soc. Am. A **30**(12), 2482 (2013). https://doi.org/10.1364/JOSAA.30.002482
13. Torrey, L., Shavlik, J.: Transfer learning. In: Handbook of Research on Machine Learning Applications and Trends: Algorithms, Methods, and Techniques, pp. 242–264. IGI global (2010)
14. Tyson, R.K.: Principles of Adaptive Optics, p. 141056. CRC Press, Boca Raton (2015)
15. Yang, H., Bharmal, N., Myers, R., Younger, E.: Laboratory demonstration of an alternative laser guide stars wavefront sensing technique—projected pupil plane pattern. J. Astron. Telescopes Instrum. Syst. **5**, 029002 (2019). https://doi.org/10.1117/1.JATIS.5.2.029002
16. Yang, H., Gonzalez Gutierrez, C., Bharmal, N.A., de Cos Juez, F.J.: Projected Pupil Plane Pattern (PPPP) with artificial neural networks. Mon. Not. RAS **487**(1), 1480–1487 (2019). https://doi.org/10.1093/mnras/stz1362

Author Index

F. Zayas-Gato et al. (Eds.): DCAI 2024, LNNS 1173, p. 95, 2024.
https://doi.org/10.1007/978-3-031-73910-1

The manufacturer's authorised representative in the EU is Springer Nature Customer Service Centre GmbH, Europaplatz 3, 69115 Heidelberg, Germany. If you have any concerns regarding our products, please contact ProductSafety@springernature.com

Printed and bound by CPI Group (UK) Ltd, Croydon, CR0 4YY
30/09/2026
02225575-0006